William Benjamin Smith

Introductory modern geometry of point, ray and circle

William Benjamin Smith

Introductory modern geometry of point, ray and circle

ISBN/EAN: 9783337281939

Printed in Europe, USA, Canada, Australia, Japan

Cover: Foto ©berggeist007 / pixelio.de

More available books at **www.hansebooks.com**

INTRODUCTORY
MODERN GEOMETRY

OF

POINT, RAY, AND CIRCLE

BY

WILLIAM BENJAMIN SMITH, A.M., Ph.D. (Goett.)

PROFESSOR OF MATHEMATICS AND ASTRONOMY
UNIVERSITY OF THE STATE OF MISSOURI

Aeltestes bewahrt mit Treue,
Freundlich aufgefaßtes Neue.
— *Goethe.*

New York
MACMILLAN & CO.
AND LONDON
1893

PREFACE.

THIS book has been written for a very practical purpose, namely, to present in simple and intelligible form a body of geometric doctrine, acquaintance with which may fairly be demanded of candidates for the Freshman class, and is in fact demanded at this University of the State of Missouri. This purpose has regulated both the amount and the character of the matter introduced. The former might have been made larger, the latter more uniform and scientific, but only — so at least it seemed to the author — at a sacrifice of usefulness under existing conditions.

Much more than one year's study can hardly be given to Plane Geometry in the majority of High Schools and Academies — a fact that sets rather narrow limits to practicable treatment of the subject. In such a course the Apollonian problem would seem to present itself as a natural and rightful goal. Besides, in its solution the logical play, while direct and simple, is yet highly instructive and even artistic, — all the concepts of the foregoing sections are summoned up and marshalled and brought to bear upon a single point. But if any such goal is to be attained in such a time, the path pursued must not be tortuous, and there will be little leisure for lateral excursions. In the Exercises, however, the view of the student is considerably widened so as to embrace most of the more familiar theorems omitted from the text. These Exercises have, in fact, been chosen with especial reference not so much to their merely disciplinary as to their didactic value, the author being persuaded that quite as good exercise may be found in going somewhither as in walking round the square. The problems proposed for solution will not merely drill the student in what he already knows, but will greatly extend his knowledge, in particular, of projection and perspective, guiding him nearly as far as he can conveniently go without the help of the Cross Ratio — a notion which the narrow scope of the work as a mere introduction seemed to exclude from employment. It is believed that advocates of the heuristic method may find in these problems ample playroom for the ingenuity of their pupils. As regards both the matter and the arrangement of this part of the book, the author would lay little claim to originality, but would rather acknowledge indebtedness to his predecessors in the attempt to modernize geometrical teaching, particularly to the valuable and indeed admirable works of Dupuis, Halsted, Henrici, Newcomb, Frischauf, Henrici and Treutlein, and Mueller.

Up to the Taction-Problem the notion of Form has dominated the whole discussion, but in the following sections certain metric relations of great importance receive due consideration.

With respect to the methods employed and the point of view assumed, a preface is no place for apology. With such as approve the resolution of the 31st Assembly of German educators:

"Im Unterricht der Elementargeometrie an Realschulen und Gymnasien bleibt die Euclidische Geometrie dem System nach bestehen, wird aber im Geiste der neueren Geometrie reformiert,"

argument would be needless; with others it might be useless. The case stands in a measure as with the Gospel saying, made

PREFACE.

for those who could receive it. The work asks to be judged, at least in its name, according to this spirit of Modern Geometry, and not according to the letter.

In the treatment of fundamental notions, of Parallels, of Proportion, and in fact throughout the book, the reader may find enough that is novel, if nothing that is new. The author cannot hope to escape criticism, and is himself aware of certain defects; but he may at least trust that his book may provoke some abler pen to more successful endeavor.

The way of Mathematics, it has been said, is broad and smooth; but it is exceeding long and exceeding steep. If the work in hand shall make the first upward steps of the climber not indeed less difficult, but quicker, longer, and less tedious, and so conserve him time, energy, and disposition for much higher ascent, there will be recompense for the labor and even for the renunciation that its preparation has entailed.

AUTHOR.

COLUMBIA, MISSOURI,
    1st October, 1892.

Articles marked with an asterisk, *, may be omitted on first reading. The early attention of the teacher is called to the Concluding Note, Arts. 353, sqq.

CONTENTS.

Art.		Page
1-159.	Linear Relations.	1-143
1- 24.	Introduction	1- 20
25, 26.	Axioms	21, 22
27- 45.	Congruence	23- 34
46- 72.	Triangles	34- 56
73- 80.	Parallelograms	56- 62
81- 89.	The 4-side, Parallels, Concurrents	63- 69
	Exercises I.	69- 76
90-107.	Symmetry	76- 90
108-139.	The Circle	90-118
140^a, 140^b.	The Circle as Envelope	119, 120
140-159.	Constructions	120-136
	Exercises II., III., IV.	136-143
160-366.	Areal Relations.	144-292
160-162.	Area	144-146
163-174.	Criteria of Equality	147-154
175-178.	Miscellaneous Applications	154-156
179-192.	Squares	157-168
193-225.	Proportion	168-187
226-234.	Similar Figures	188-192
235-280.	Constructions	192-212
281-297.	The Taction-Problem	212-225
298-328.	Metric Geometry	226-248
329-333.	Measurement of the Circle	248-253
334-336.	Measurement of Angles	253-255
337-343.	The Euclidian Doctrine of Proportion	255-261
344-353.	Maxima and Minima	261-268
354-366.	Concluding Note	268-274
	Exercises V.	274-292
	Index	293-297

GEOMETRY.

INTRODUCTION.

1. Geometry is the Doctrine of Space.
What is Space? On opening our eyes we see objects around us in endless number and variety: the book here, the table there, the tree yonder. This vision of a world outside of us is quite involuntary — we cannot prevent it, nor modify it in any way; it is called the *Intuition* (or Perception or Envisagement) *of Space*. Two objects precisely alike, as two copies of this book, so as to be indistinguishable in every other respect, yet are not the same, because they differ in place, in their positions in Space: the one is here, the other is not here, but there. In between and all about these objects that thus differ in place, there lies before us an apparently unoccupied region, where it seems that nothing *is*, but where anything *might be*. We may imagine or suppose all these objects to vanish or to fade away, but we cannot imagine this region, either where they were or where they were not, to vanish or to change in any way. This region, whether occupied or unoccupied, where all these objects are and where countless others might be, is called **Space**.

2. There are certain *elementary facts*, that is, facts that cannot be resolved into any simpler facts, about this Space, and these deserve special notice.

A. Space is fixed, permanent, unchangeable. The objects in Space, called bodies, change place, or may be imagined to change place, in all sorts of ways, without in the least affecting Space itself. Animals move, that is, change their places, hither and thither; clouds form and transform themselves, drifting before the wind, or dissolve, disappearing altogether; the stars circle eternally about the pole of the heavens; sun, moon, and planets wander round among the stars; but the blue dome of the sky,* the immeasurable expanse in which all these motions go on, remains unmoved and immovable, as a whole and in all its parts, absolutely the same yesterday, to-day, and forever.

B. Space is **homœoidal**; *i.e.* it is precisely alike throughout its whole extent. Any body may just as well be here, there, or yonder, so far as Space is concerned. A mere change of place in nowise affects the Space in which the change, or motion, occurs.

C. Space is **boundless**. It has no beginning and no end. We may imagine a piece of Space cut out and colored (to distinguish it from the rest of Space); the piece will be bounded, but Space itself will remain unbounded.

N. B. When we say that Space is unbounded, we do *not* mean that it is *infinite*. Suppose an earthquake to sink all the land beneath the level of the sea, and suppose this latter at rest; then its outside would be unbounded, without beginning and without end, — a fish might swim about on it in any way forever, without stop or stay of any kind. But it would *not* be infinite; there would be exactly so many square feet of it, a finite number, neither more nor less. Likewise, the fact that bodies may and do move about in space every way without let or hindrance of any kind implies

---

* Appearing blue because of the refraction of light in the air.

that Space is boundless, but by no means that it is infinite. For all we know there may be just so many cubic feet of Space; it may be just so many times as large as the sun, neither more nor less. This distinction between unbounded and infinite, first clearly drawn by Riemann, is fundamental.

D. Space is **continuous**. There are no gaps nor holes in it, where it would be impossible for a body to be. A body may move about in Space anywhere and everywhere, ever so much or ever so little. Space is itself simply where a body may be, and a body may be anywhere.

E. Space is **triply extended**, or has **three dimensions**. This important fact needs careful explication.

In telling the size of a box or a beam we find it necessary and sufficient to tell three things about it: its length, its breadth, and its thickness. These are called its dimensions; knowing them, we know the size completely. But to tell the size of a ball it is enough to tell one thing about it, namely, its diameter; while to tell the size of a chair we should have to tell many things about it, and we should be puzzled to say what was its length, or breadth, or thickness. Nevertheless, it remains true that Space and all bodies in Space have just three dimensions, but in the sense now to be made clear.

We learn in Geography that, in order to tell accurately where a place is on the outside of the earth, which may conveniently be thought as a level sheet of water, it is necessary and sufficient to tell *two* things about it; namely, its latitude and its longitude. Many places have the same latitude, and many the same longitude; but no two have the same latitude *and* the same longitude. It is not sufficient, however, if we wish to tell exactly where a thing is in Space, to tell two things about it. Thus, at this moment the bright star Jupiter is shining exactly in the south; we also know its altitude, how high it is above the horizon (this altitude is measured *angularly* — a term to be explained hereafter, but with which we have no present concern). But the knowledge of these two facts merely enables us to *point towards* Jupiter; they do not fix his place definitely, they do not say how far away he is: we should point towards him the same way whether he were a mile or a million of miles distant. Accordingly, a third thing must be known about him, in order to know precisely where he is; namely, his distance from us. But when this third thing is known, no further knowledge about his place is either necessary or possible. Once more, here is the point of a pin. Where is it in this room? It is five feet above the floor. This is not enough, however, for there are many places five feet above the floor. It is also ten feet from the south wall, but there are yet many positions five feet from the floor and ten feet from the south wall, as we may see by slipping a cane five feet long sharpened to a point, upright on the floor, keeping the point always ten feet from the south wall. But as it is thus slipped along, the point of the cane will come to the point of the pin and then will be exactly twelve feet from the west wall. If it now move ever so little either way east or west, it will no longer be at the pin-point and no longer twelve feet from the west wall. So there is one, and only one, point that is five feet from the floor, ten feet from the south wall, and twelve feet from the west wall. Hence it is seen that these three facts fix the position of the pin-point exactly. A fourth statement, as that the point is nine from the ceiling, will either be superfluous, if the ceiling is fourteen feet high, being implied in what is already said, or else incorrect, if the ceiling is not fourteen feet high, contradicting what is already said. In general, with respect to any position in Space it is *necessary* to know three independent facts (or data), and it is *impossible* to know any more. All other knowledge about the position is involved in this knowledge, which is necessary and sufficient to enable us to answer any rational question that can be put with respect to the position. Accordingly, since any position in Space is known completely when, and only when, *three* independent data are known about it, we say that Space is *triply* or *threefold extended*, or has *three dimensions*. The dimensions are any three independent things that it is necessary and sufficient to know about any position in Space, as of the pinpoint or of Jupiter, in order to know exactly where it is.

3. But with respect to the outside of the earth, viewed as a level sheet of water, we have seen that only two data, as of latitude and longitude, are necessary and ·sufficient to fix any position on it; neither are more than two independent data possible; all other knowledge about the position is involved in the knowledge of these two data about it. Accordingly we say of such outside of the earth that it is *doubly* or *two-fold extended*, is bi-dimensional, or has two dimensions; and we name every such *outside*, every such bi-dimensional region, a **surface**. Such is the top of the table : to know where a spot is on it we need know two, and only two, independent facts about it, as how far it is from the one edge and how far from the other. (Which other? and why?)

We see at once that a surface is no part of Space, but is only a *border* (doubly extended) between two parts of Space. Thus, the whole earth-surface is no part either of the earth-space or of the air-space around the earth, but is the boundary between them. A soap-bubble floating in the air is not a surface ; though exceedingly thin, it has some thickness and occupies a part of space ; the outside of the film is a surface, and so is the inside, and these are kept apart by the film itself. If the film had no thickness, the outside and the inside would fall together, and the film would be a surface; namely, the outside of the Space within and the inside of the Space without.

4. Consider now once more this earth-surface, still viewed as a smooth level sheet of water. From Geography we learn that there are two extreme positions on this surface that are called *poles* and that do not move at all as the earth spins round on her axis. We also learn that there is a certain region of positions just midway between these poles and called the **Equator**. This Equator is no part of the surface; it is only a border or boundary between two parts of the globe-surface, which are called *hemispheres*. To know where any position is on this border, it is necessary and sufficient to know *one* thing about it, namely, its longitude; neither is any other independent knowledge about the position possible; all other knowledge is involved in this one knowledge. Accordingly we say of this border, the Equator, that it is *simply extended*, or has one dimension only. Every such one-dimensional border is called a **line**, and its one dimension is named *length*. A line, then, has length, but neither breadth nor thickness.

5. Lastly, consider a part of a line, as of the Equator, say between longitudes 40° and 50°. The ends of this part bound it off from the rest of the equator, but they themselves form no part of the Equator. They are called **points**; they have position merely, but no extent of any kind, neither length nor breadth nor thickness, — they are wholly *non-dimensional*.

INTRODUCTION. 7

*6. It is noteworthy that extents, or regions, are bounded by extents of fewer dimensions, and themselves bound extents of more dimensions. Thus, lines are bounded by points, and themselves bound surfaces; surfaces are bounded by lines, and themselves bound spaces; spaces are bounded by surfaces, and themselves bound — what? If anything at all, it must be some extent of still higher order, of *four* dimensions. But here it is that our intuition fails us; our vision of the world knows nothing of any fourth dimension, but is confined to three dimensions. If there be any such fourth dimension, we can know nothing of it by intuition; we cannot imagine it. In music, however, we do recognize four dimensions: in order to know a note completely, to distinguish it from every other note, we must know four things about it: its pitch, its intensity, its length, its timbre, — how high it is, how loud it is, how long it is, how rich it is. While, then, extents of higher dimensions may be unimaginable, they are not at all unreasonable.

This doctrine of dimensions is of prime importance, but rather subtile; let not the student be disheartened, if at first he fail to master it.

6. We may see and handle bodies, which occupy portions of Space; but not so surfaces, lines, points, which occupy no Space, but are merely regions in Space. Here we must invoke the help of the logical process called **abstraction,** *i.e.* withdrawing attention from certain matters, disregarding them, while regarding others. A sheet of paper is not a surface, but a body occupying Space. However thin, it yet has some thickness. But in thinking about it we may leave its thickness out of our thoughts, disregard its thickness altogether; so it becomes for our thought, though not for our senses or imagination, a *surface*. The like may be

8 GEOMETRY.

said of the film of the soap-bubble. Again, consider the pointer. It is a body or solid, not only long, but wide and thick; it occupies Space. It is neither line nor surface. But we may, and do often, disregard wholly two of its dimensions, and attend solely to the fact that it is *long*. Thus it becomes for our thought a *line*, though not for our senses or imagination. So the mark made with chalk or ink or pencil is a body, triply extended; but we disregard all but its length, and it becomes for our reason a *line*. Lastly, we make a dot with pen or chalk or pencil; it is a body, tri-dimensional, occupying Space. But we may disregard all its dimensions, and attend solely to the fact that it has position, that it is here, and not there. So it becomes in our thought a *point*. By such abstraction the earth, the sun, the stars, the planets, may all be treated as points.

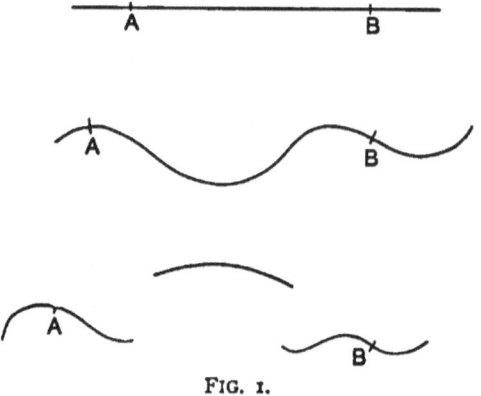

FIG. 1.

7. Inasmuch as Space is continuous, there may also be continuous surfaces and lines; and the only surfaces and lines treated in this book are continuous, without holes, gaps, rents, breaks, or interruptions of any kind in their extent.

It is important to note that in passing from any position *A*

INTRODUCTION. 9

to another B on a continuous line, a moving point P must pass through a complete series of intermediate positions; *i.e.* there is no position on the line between A and B that the point P would not assume in going from A to B. (Fig. 1.)

*8. Starting from the notion of Space, we have attained the notions of surface, line, and point, in two ways: by treating them as borders, and by the process of abstraction. But we may reverse this order and attain the notions of line, surface, and solid or space from the notion of point, with the help of the notion of *motion*, thus: Let a *point* be defined as having *position without parts or magnitude* of any kind. Let it move continuously through Space from the position A to the position B. To know where it is at any stage of its motion along any definite path, it is necessary

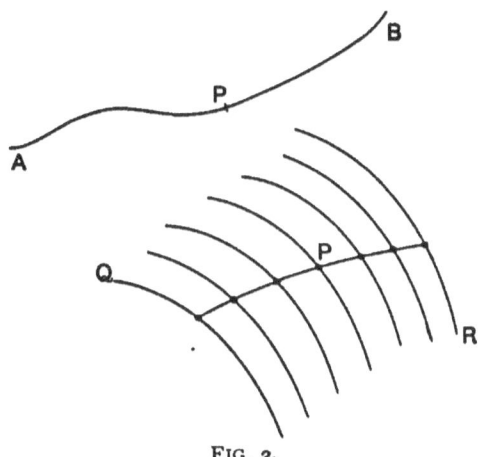

FIG. 2.

and sufficient to know one thing; namely, how far it is from A. Hence its path is a *one*-dimensional extent, or what we call a *line*.

Now let a line move in any definite way from any position Q to any other position R. To know the position of any point of its path, it is necessary and sufficient to know two things; namely, the position of the point on the moving line and the position of the moving line itself: hence the path of the line is a *two*-dimensional extent, which we have already named a *surface*. (Fig. 2.)

Now let a surface move in any definite way from any position U to any other position V. To know the position of any point on its path it is necessary and sufficient to know three things about it; namely, its position on the moving surface (which, we know, counts as two things) and the position of the moving surface itself. Hence the path of the surface is a *three*-dimensional extent, which we have already named a *solid* or a part of *space*.

Now, if we let a solid move, what will its path be? Naturally we should expect it to be a *four*-dimensional extent, but no such extent is yielded in our experience by any motion of a solid — the path of a solid is nothing but a solid. The explanation of the apparent inconsistency is very simple, to-wit: A piece of a line traces out a surface only when it moves out from the line itself, — if one part were to slip round on another part of the same line, it would trace out no surface at all as its path; likewise, a piece of a surface traces out a solid as its path only by moving out from the surface itself, — if one part were to slip round on another part of the surface, it would trace out no solid at all as its path. So, if a piece of our space could move out from space itself, it would trace out a four-fold extent as its path; in fact, however, no part of space can move out from space; on the contrary, it can only slip along *in* space, from one part of space to another, and hence does not trace out any four-fold extended path.

INTRODUCTION. 11

9. Space, we have seen, is homœoidal, everywhere alike. We naturally inquire : Is there any homœoidal surface? In general, surfaces are certainly *not* homœoidal. Consider an egg-shell, and by abstraction treat it as a surface. It is not alike throughout; the ends are not like each other, and neither is like the middle region. Suppose a piece cut out anywhere; if slipped about over the rest of the shell, this piece will *not fit.* But now consider a smooth round ball covered with a thin rigid film, and treat this film as a surface, by disregarding its thickness. Suppose a piece of the film cut out and slipped round over the rest of the film : the piece will fit everywhere perfectly, the surface is homœoidal; it is called a **sphere**-surface.

N.B. The precise definition of this surface is that *all its points are equidistant from a point within, called the centre.* Suppose a rigid bar of any shape, pointed at both ends, and movable about one end fixed at a point; then the other end will move always on a sphere-surface, which is *the whole region where the moving end may be.* Since Space is homœoidal around the fixed point, the surface everywhere equidistant from the point is also homœoidal.

Now turn over the piece cut out of this spherical film and slip it about the film : it no longer fits anywhere at all — the surface is homœoidal, but *not reversible.*

10. But now consider a fine mirror covered with a delicate film, which by abstraction we treat as a surface. Suppose a piece cut out of the film and slipped about over it : the piece fits everywhere ; turn it over, re-apply it, and slip it about : it still fits everywhere — the surface is both *homœoidal* and *reversible;* it is called a **plane**-surface.

* N.B. A precise definition of this surface is the following : Take two points *A* and *B* and suppose two equal spherical bubbles formed about A and B as centres. Let them expand, always equal to each other, until they meet, and still keep on expanding. The line where the equal (Fig. 3) spherical bubbles, regarded as surfaces, meet, has all its points just as far from A as from B. As the bubbles still expand, this line, with all its points equidistant from A and B, itself expands and traces out a *plane* as its path through Space.

$\dot{A} \qquad \dot{B}$
FIG. 3.

Hence we may define the *plane* as the region (or surface) where a point may be that is equidistant from two fixed points. Instead of region it is common to say **locus**, *i.e.* place. Briefly, then, a *plane is the locus of a point equidistant from two fixed points*. It is evident that the plane, as thus defined, is reversible; for since the bubbles about A and B are all the time precisely equal, to exchange A and B, or to exchange the sides of the plane, will make no difference whatever. Thus the plane cuts the Space evenly half in two; and since Space itself is homœoidal, so also is this section or surface that halves it exactly. The superiority of this definition consists in its not only telling what surface the plane is, but also making clear that there actually *is* such a surface.

11. The mirror is the nearest approach that we can make to a perfect plane surface; the blackboard is not plane, it is rough and warped; but we shall disregard all its unevenness and treat it as a plane extended through Space without end. Any surface may be dealt with as a plane by abstraction, being thought as *homœoidal* and *reversible*.

12. On this board, regarded as a plane, we draw a chalk-mark, abstract from all its dimensions but its length, and

treat it as a line. This line is plainly not alike throughout; a piece cut out and slipped along it will not fit (Fig. 4).

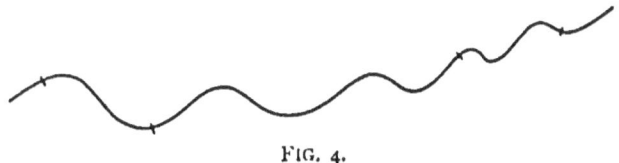

FIG. 4.

But here is a line homœoidal, alike in all its parts; it is drawn with a pair of compasses and is called a **circle** (Fig. 5). One point of the compasses is held fast at the *centre O*, while the other traces out the circle as its path in

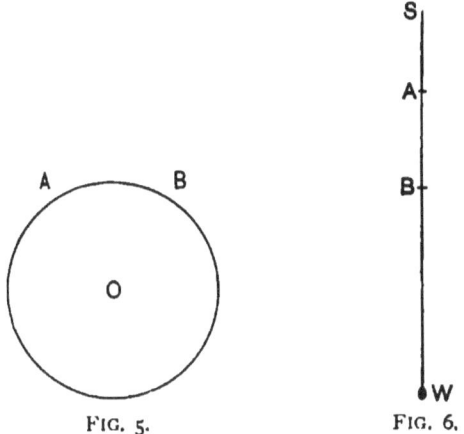

FIG. 5.   FIG. 6.

the plane. The circle is *the locus of a point in the plane equidistant from a fixed point.* Since the plane is homœoidal, so too is this circle (see Art. 10); a piece, called an **arc**, cut out and slipped round will everywhere fit on the circle. But turn it over and slip it round, — it fits nowhere; the circle is *not* reversible. It divides the plane into two parts, not halves, that are not alike along the dividing line. But now suppose a perfectly flexible string fastened at S and stretched by a weight W. Its length only being regarded, it is a line homœoidal, alike throughout, and also reversible; any part AB will not only fit perfectly anywhere on it, but will also fit when reversed, turned end for end. Such a line is called *right*, or *straight*, or *direct*, or a **ray**. Extended indefinitely, it cuts the whole plane into two halves precisely alike along the ray itself.

*N.B. The common line where the two spherical bubbles of Art. 10 meet is a circle, for it is plainly precisely alike all around; it is homœoidal, being the intersection of two homœoidal surfaces, namely, the two equal sphere-surfaces; it is also in a plane, and in fact traces out the plane by its expansion as the bubbles expand.

To get accurately the notion of the **ray** or *straight line*, we need another point C, and a third expanding bubble always equal to those about A and B. The circular intersection of the bubbles about A and B will trace out one plane; of those about B and C will trace out another plane; of those about C and A will trace a third plane. All the points where the first two planes intersect will be equidistant from A and B and C, and no other points will be; the same may be said of all points where the second and third planes meet, and of all points where the third and first meet; hence all three of the planes meet together, and they meet only together. Also, the line where they meet has every one of its points equidistant from all the three points, A, B, C; hence it is the **locus** *of a point equidistant from three fixed points*. Moreover, it is *homœoidal* and *reversible*, since it is the intersection of two planes, which are homœoidal and reversible; hence it is what we call a straight line, or right line, or *ray*.

13. We may now define:

A **sphere**-surface is the locus of a point at a fixed distance from a fixed point. It is homœoidal, but not reversible.

A **plane** is the locus of a point equidistant from two fixed points. It is both homœoidal and reversible.

A **ray** is the locus of a point equidistant from three fixed points. It is both homœoidal and reversible; it is also the intersection of two planes.

A **circle** is the locus (or path) of a point in a plane at a fixed distance from a fixed point. It is homœoidal, but not reversible. It is also the locus (or path) of a point in space at a fixed distance from two points; it is also the intersection of two equal sphere-surfaces.*

14. It is only with the foregoing figures and combinations of them that we have to deal in this book. Circles and rays may be drawn with exceeding accuracy, but any lines, however roughly drawn, may answer our logical purposes as well as the most accurately drawn; we have only, by abstraction, to treat them as having the character of the lines in question.

Circles and sphere-surfaces are unbounded, without beginning or end, but both are finite: we shall learn how to measure them.

---

* In the foregoing free use has been made of the notion of *equidistance* without formal definition, because of its familiarity. We may, however, say precisely: If A and B be two points, the ends of a rigid bar of any shape, and if A be held fast, then all the points on which B can fall are equidistant from A, and no other points are equidistant with them. They all lie on a closed surface, called a sphere-surface. All points within this surface are said to be *less* distant, and all points without are said to be *more* distant, from A than B is. Herewith, then, we tell exactly what we mean by equidistant, less distant, and more distant, but we make no attempt to define *distance* in general, which is difficult and unnecessary to our purpose.

15. Any geometric element or combination of geometric elements, as points, lines, surfaces, is called a geometric **figure**. It is a fundamental assumption, justified by experience, that space is homœoidal, that figures or bodies are not affected in size or shape by change of place. Two figures that may be fitted exactly on each other, or may be thought so fitted, are called **congruent**. Any two points, lines, or parts of the two figures, that fall upon each other in this superposition are said to **correspond**. It is manifest that all planes are congruent and all rays are congruent. Rays and planes are unbounded, but whether or not they are finite is a question that we are unable to answer.

16. Any part of a circle or ray, as AB, is bounded by two end-points, A and B, and is finite; the one is named an **arc** (Fig. 5), the other a **tract**, *sect*, or *line-segment*. Each is denoted by the two letters denoting the ends, as the tract AB, the arc AB. Sometimes it is important to distinguish these end-points as beginning and end proper; we do this by writing the letter at the beginning first.

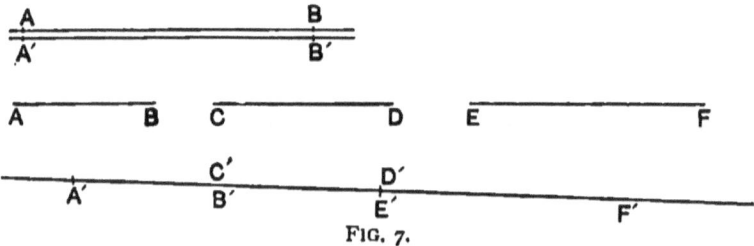

FIG. 7.

17. Two tracts, AB and $A'B'$, are called equal when the end-points of the one may be (Fig. 7) simultaneously fitted on the end-points of the other.

If we have a number of tracts, AB, CD, EF, etc., and we lay off successively on a ray tracts $A'B'$, $C'D'$, $E'F'$, etc.,

INTRODUCTION. 17

respectively equal to *AB, CD, EF*, etc., the end of the first being the beginning of the second, and so on, while no part of one falls on any part of another, we are said to *add* or **sum** the tracts *AB*, etc. Each is called an *addend* or **summand**, and the whole tract from first beginning to last end is called the **sum**.

Equality is denoted by the bars (=) between the equals, as $AB = CD$.

18. If, when the beginning A is placed on the beginning C, the end B does not fall on the end D, the tracts are unequal, and we write $AB \neq CD$. If B falls between C and D, then AB is called less than CD, $AB < CD$; but if D falls between A and B, then AB is called greater than CD, $AB > CD$. In either case, the tract BD or DB, between the two ends of the tracts, whose beginnings coincide, is called the **difference** of the two tracts, and we are said to **subtract** the one from the other. Ordinarily we mention the greater tract first in speaking of difference.

19. The symbols of addition and subtraction are + and — (plus and minus), thus:

$$AB + CD = AD \text{ and } AB - CD = BD.$$

It is important to note here the order of the letters. In summing a number of tracts, as *AB, CD, EF*, etc., to *KL*,

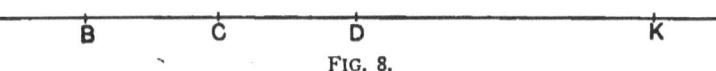
FIG. 8.

we have $AB + CD + EF \cdots + KL = AL$ (Fig. 8). The order of the summands is *indifferent*, and this important fact is called the **Commutative Law** of Addition. Thus

$$AB + CD + EF = AB + EF + CD = EF + AB + CD, \text{ etc.}$$

GEOMETRY.

20. When beginning and end of a tract or of any magnitude are exchanged, the tract or magnitude is said to be **reversed**, and the reverse is denoted by the sign —. Thus the reverse of AB is BA, or $AB = -BA$. If we add a magnitude and its reverse, the sum is 0, or

$$AB + (-AB) = AB + BA = 0.$$

The same result 0 is obtained by subtracting, from a magnitude, itself or an equal magnitude; and, in general, it is plain that to subtract CD yields the same result as to add (Fig. 9) the reverse DC. The reverse of a magnitude is

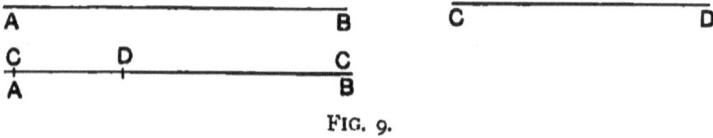

FIG. 9.

often called its *negative*, the magnitude itself being called its *positive*.

Similar rules hold for adding and subtracting arcs of a circle or of equal circles.

ANGLES.

21. The indefinite extent of a ray on one side of a point O, as OA, is called a **half-ray**: it has a beginning O, but no end. Two half-rays, OA and OA', which together make up a whole ray, are called *opposite* or **counter** (Fig. 10).

Now let two half-rays, OA and OB, have the same beginning O; the opening or spread between them is a magnitude: it may be greater or less. Suppose OA and OB to be two very fine needles pivoted at O; then OB may fall exactly on OA, or it may be turned round from OA; and

INTRODUCTION. 19

the amount of turning from OA to OB, or the spread between the half-rays, is called the **angle** between them. We may denote it by a Greek letter, as a, written in it; or by a large Roman letter, as O, at its vertex (where the half-rays meet); or by three such letters, as AOB, the middle

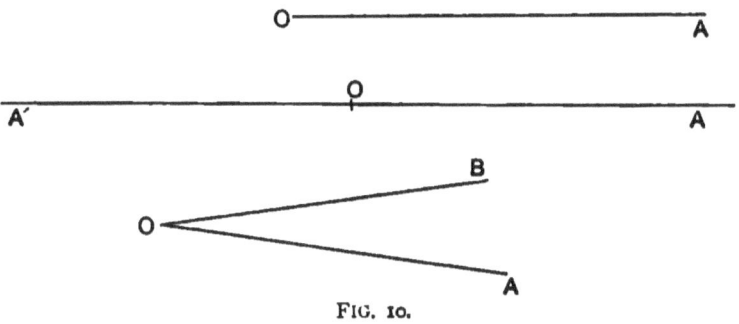

FIG. 10.

one being at the vertex, the other two anywhere on the half-rays. The symbol for angle is ⅄.

22. The angle is perfectly definite in size, it has two *ends* or boundaries; namely, the two half-rays, sometimes called **arms**. When we would distinguish these arms as beginning and end, we mention the letter on the beginning-arm first, and the letter on the end-arm last; thus, AOB; here OA is the *beginning* and OB the *end* of the angle.

Exchanging beginning and end reverses the angle; thus, $BOA = -AOB$.

23. Two angles whose ends or arms may be made to fit on each other simultaneously are named **equal**; they are also *congruent*. Two angles whose arms will not fit on each other simultaneously are *unequal;* and that is the less angle whose end-arm falls *within* the other angle when their beginnings

20   GEOMETRY.

coincide; the other is the greater; thus, $AOB > AOC$ (Fig. 11).

24. We sum angles precisely as we sum tracts; we lay off α, β, etc., around O, making the end of each the beginning of the next: the angle from first beginning to last end

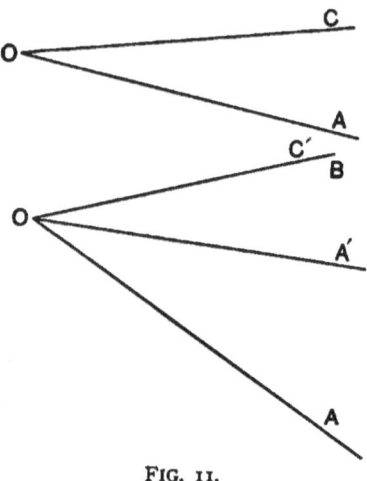

FIG. 11.

is the *sum*. So, too, in order to subtract β from α, lay off β from the beginning towards the end of α; the angle from the end of β to the end of α is the difference, $\alpha - \beta$. Or we may add to α the reverse (or negative) of β: the sum will be $\alpha + (-\beta)$ or $\alpha - \beta$ (Fig. 12).[1]

---

[1] It is important to note the close correspondence of tract and angle: the former is related to points as the latter is to rays (or half-rays). The tract is the simplest magnitude that lies between points, that distinguishes them and keeps them apart; likewise the angle is the simplest magnitude that lies between rays (in a plane), that distinguishes them and keeps them apart. So, too, we define equality and inequality among tracts and among angles, quite similarly, and without being compelled beforehand to form the notion of the size either of a tract or of an angle. We may now define the *distance* between two points to be the *tract* between them, and the *distance* between two (half-)rays to be the *angle* between them, leaving for future decision which tract and which angle if there should prove to be several.

INTRODUCTION. 21

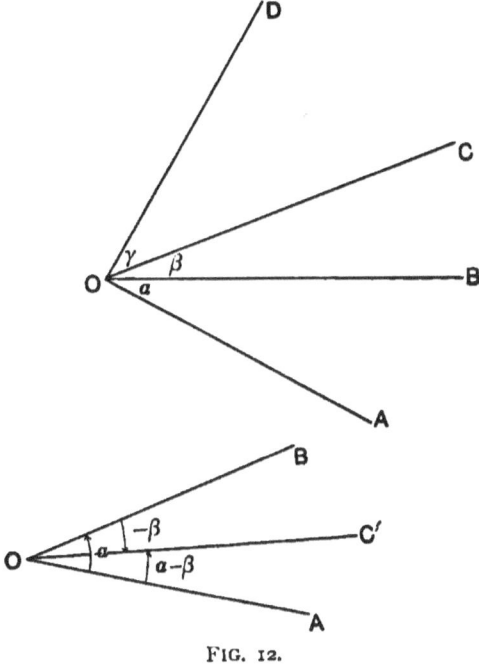

FIG. 12.

AXIOMS.

25. At this stage we must recognize and use certain dictates or irresoluble facts of experience, called **axioms**. ('Αξίωμα means *something worthy*, like the Latin *dignitas;* in fact, older writers use *dignity* in the sense of *axiom*. But Euclid's phrase is κοιναι ἐννοιαι = *common notions.*) Some have no special reference to Geometry, but pervade all of our thinking about magnitudes; such are

(1) Things equal to the same thing are equal to each other.

(2) If equals be added to, subtracted from, multiplied by, or divided by, equals, the results will be equal.

(3) If equals be added to or subtracted from unequals, the latter will remain unequal as before.

(4) The whole equals, or is the sum of, all its distinct parts, and is greater than any of its parts.

(5) If a necessary consequence of any supposition is false, the supposition itself is false.

Others concern Geometry especially, as :

(6) All planes are congruent.

(7) Two rays can meet in only one point.

The extremely important axiom (7) may be stated in other equivalent ways, thus : Two rays cannot meet in two or more points; or, Two rays cannot have two or more points in common; or, Only one ray can go through two fixed points; or, A ray is fixed by two points.

26. A statement or declaration in words is called a *proposition*. The propositions with which we have to deal state geometric facts and are also called **Theorems** (θεωρημα, from θεωρειν, *to look at*, means the *product of mental contemplation*). Propositions are often incorrect; theorems, never. Subordinate facts, special cases of general facts, and facts immediately evidenced from some preceding facts, are called **Corollaries** or Porisms (πορισμα = *deduction*).

We may now proceed to investigate lines and angles, and find out what we can about them. The first and simplest things we can learn concern

CONGRUENCE.

27. Theorem I. — *All rays are congruent.*

Proof. Let L and L' be any two rays (Fig. 13). On L take any two points, A and B; on L' take any two points, A' and B', so that the tract AB shall equal the tract $A'B'$. Think of L and L' as extremely fine rigid spider-threads, and in thought place the ends of the tract AB on the ends of the tract $A'B'$, A on A', and B on B'.

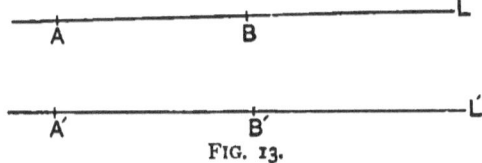

FIG. 13.

Then A and A' become one and the same point, and B and B' become one and the same point; through these two points only one ray can pass (by Axiom 7): hence L and L', which go through these two points, now become one and the same ray; that is, they fit precisely, they are congruent. *Quod erat demonstrandum* = *which was to be proved* = ὅπερ ἔδει δεῖξαι, — the solemn Greek formula; whereas the Hindu, appealing directly to intuition, merely said *Paçya* — Behold !

28. In the foregoing proof we assumed that on any ray we could lay off a tract equal to a given tract, or that on any ray we could find two points, A and B, as far apart as two other points, A' and B'. This assumption that something can be done, is called a **Postulate** (αἴτημα), *i.e.* a *demand*, which must be granted before we can proceed further. Actually to carry out the construction, we need a pair of compasses.

29. Theorem II. — *If two points of a ray lie in a certain plane, all points of the ray lie in that plane.*

Proof. Regard the surface of paper or of the blackboard as a plane, and suppose it covered with a fine rigid film, itself a plane. Let L be any ray having two points, A and B, in this plane. Through these two points suppose a second plane drawn or passed; by definition (Art. 13) it will intersect our first plane, or film, along a ray I; this ray I goes through the two points, A and B, and lies wholly (with all its points) in the first plane; also the ray L goes through A and B, and only one ray can go through the same two points, A and B, by Axiom 7; hence L and I are the same ray; but I has all its points in the first plane; hence L has all its points in the first plane. Q. E. D.

Query: What *postulate* is assumed in this proof?

Corollary. If a ray turn about a fixed point P, and glide along a fixed ray L, it will trace out a plane (Fig. 14).

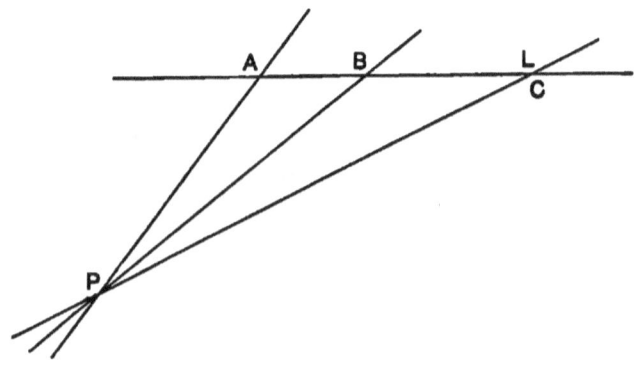

FIG. 14.

For it will always have two points — namely, the fixed point and a point on the fixed ray — in the plane drawn through the fixed point and the fixed ray.

Query: What postulate is here implied?— Henceforth it is understood that all our points, lines, etc., are **complanar**, *i.e.* lie in one and the same plane.

30. In the foregoing Theorem and Corollary we observe clauses introduced by the word if. Such a clause is called an **Hypothesis**, *i.e.* a supposition. The result reached by reasoning from the hypothesis and stated immediately after the hypothesis, is called the **Conclusion**.

31. All logical processes consist in one or both of two things: the formation of concepts, as of lines, surfaces, angles, etc., and the combination of these concepts into propositions. Geometric concepts are remarkable for their perfect clearness and precision — we know exactly what we mean by them; this cannot be said of many other concepts, about which diverse opinions prevail, as in Political Economy. Hence it is that Geometry offers an unequalled gymnasium for the reason or logical faculty. We shall now generate some new concepts. Let the student note their definiteness as well as the mode of their formation.

32. Let OA and OB be any two co-initial half-rays, forming the angle AOB. Think of OA as held fast and of OB as turning about the pivot O, starting from the position OA. As it turns (counter-clockwise), the (Fig. 15) angle

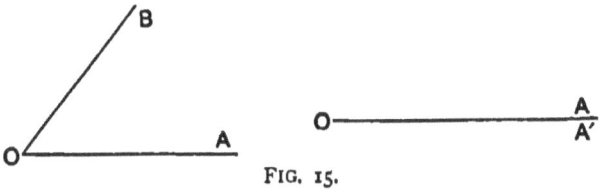

FIG. 15.

AOB increases. Finally, let it return to its original position, OA; then the whole amount of turning from the upper side of OA back to the under side of OA, or the full spread around the point O, is called a *full* angle (or **round** angle, or *circum*-angle, or perigon). Think of a fan opened until the first rib falls on the last. — Note that the upper and under sides of OA are exactly the same in position, and are distinguished only in thought. (Think of a circular piece of paper slit straight through from the edge to the centre.) The like may be said of the two sides of any line or surface. We can now prove

33. Theorem III. — *All round angles are congruent.*

Proof. Let AOB and $A'O'B'$ (Fig. 16) be any two round angles. Slip the half-ray OA down, and turn it till OA falls on $O'A'$; they will fit perfectly (why?); the

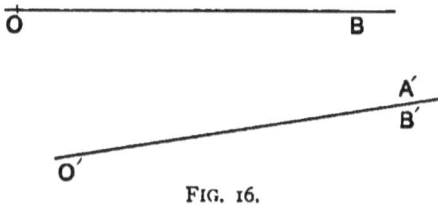

FIG. 16.

whole round angle about O will fit perfectly on the whole round angle about O' (why?); hence the two full angles are congruent. Q. E. D.

N.B. In this slipping of figures about in the plane, it is well to imagine the plane to consist of two very thin, perfectly rigid, smooth and transparent films; also, to imagine one figure drawn in the lower film and one in the upper; and to imagine the upper slipped about at will over the lower.

Query: On what cardinal property of the plane do these considerations hinge?

TH. IV.] CONGRUENCE. 27

34. From O draw any half-ray OA; then any second half-ray from O, as OB, will (Fig. 17) cut the round angle AOA into two angles, AOB and BOA. The end OB of the first falls on the beginning, OB, of the second; while

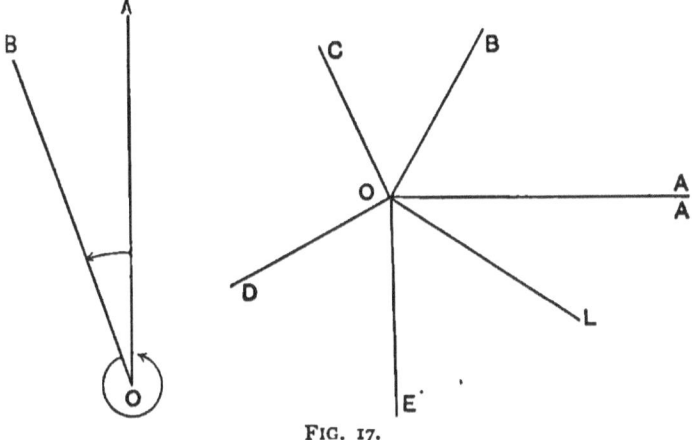

FIG. 17.

the end, OA, of the second falls on the beginning, OA, of the first. Hence the round angle AOA is their sum, by Art. 24.

If we draw any number of half-rays, OB, OC, etc., $\cdots OL$, the round angle will still be the sum of the consecutive angles AOB, BOC, etc., $\cdots LOA$; hence we discover and enounce this

Theorem IV. — *The sum of the consecutive angles about a point in a plane is a round angle.*

N.B. We cannot apply Axiom 1 immediately, because we do not know, except by Art. 24, what is meant by a *sum* of angles.

35. In the foregoing article we have exemplified the *erotetic*, questioning, investigative method, in which the result

is not announced until it is actually discovered and established. In Theorems I., II., III., on the other hand, the *dogmatic* procedure was illustrated, the fact or proposition being announced beforehand, while the demonstration followed after. Each method has its merits, and we shall employ both.

36. As OB turns round from the upper to the under side of OA, the angle AOB begins by being less than BOA and ends by (Fig. 19) being greater than BOA. The plane

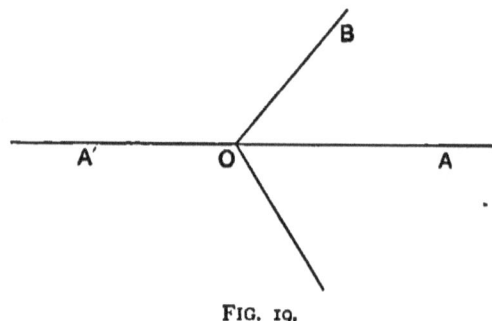

FIG. 19.

is continuous, the turning is continuous, the change in size is continuous; hence, in passing from the stage of being *less* to the stage of being *greater*, the angle has passed through the intermediate stage of being *equal;* let OA' be the position of the rotating half-ray at this stage of equality, then $AOA' = A'OA$. Two equal parts making up a whole are called **halves**; hence AOA' and $A'OA$ are halves of the full angle AOA; they are named **straight** (or flat) angles.

37. Now, — Halves of equals are equal;
All straight angles are halves of equals (namely, equal round angles);

Hence

Theorem V. — *All straight angles are equal.*

This argument here given *in extenso* is a specimen of a syllogism (συλλογισμος = computation = thinking together). The first two propositions are called **premisses**, the third and last, in which the other two are thought together, is called **conclusion**. All reasoning may be syllogized, but this is rarely done, as being too formal and tedious.

38. **Theorem VI.** — *Two counter half-rays bound a straight angle.*

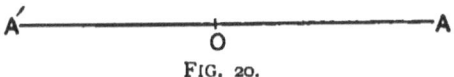

FIG. 20.

For, let OA and OA' be two such counter half-rays (Fig. 20) forming the whole ray AA'. Turn the upper half of the plane film round O as pivot until the upper OA' falls on the lower OA; then, since the ray is reversible, the ray AA' will fit exactly on the ray $A'A$; *i.e.* the two angles AOA' and $A'OA$ are congruent and equal; and the two compose the round angle AOA; hence each is half of AOA; *i.e.* each is a straight angle. Q. E. D.

39. **Theorem VII.** — Conversely, *The half-rays bounding a straight angle are counter.*

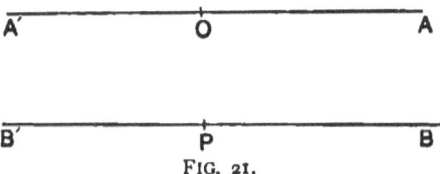

FIG. 21.

Let OA and OA' bound a straight angle (Fig. 21) AOA'; also let PB and PB' be two counter half-rays; then they bound a straight angle BPB', by Theorem VI. Since all straight angles are congruent, we may fit these two on each other; *i.e.* we may fit OA and OA' on PB and PB'; but BB' is a ray; so then is AA'; *i.e.* OA and OA' are counter. Q. E. D.

40. We may *define* a straight angle as *an angle bounded by counter half-rays*. Then we may prove Theorem V. thus:

The ends of all straight angles are pairs of counter half-rays (or form whole rays);

But all such pairs (or whole rays) are congruent (by Theorem I.);

Therefore, all ends of straight angles are simultaneously congruent.

But when the ends of angles are (simultaneously) congruent, so are the angles themselves.

Hence all straight angles are congruent. Q. E. D.

Here the first conclusion, introduced by "therefore," is deduced from two premisses; but the second, introduced by "hence," is apparently deduced from only one. Only apparently, however; for one premiss was understood but not expressed; namely, *all straight angles are angles whose ends are congruent*. Without some such implied additional premiss, it would be impossible to draw the conclusion. Such a maimed syllogism, with only one expressed premiss, is called an **enthymeme**. The great body of our reasoning is enthymematic. We shall frequently call for the suppressed premiss or *reason* by a parenthetic question (Why?).

41. Now draw two rays, LL' and MM', meeting at O. Each divides the round angle about O into two equal straight angles, and together they (Fig. 22) form four angles α, β, α', β'. Two angles, as α and β, that have a common arm, are called **adjacent**. Accordingly we see at once:

Theorem VIII. — *Where two rays intersect, the sum of two adjacent angles is a straight angle.*

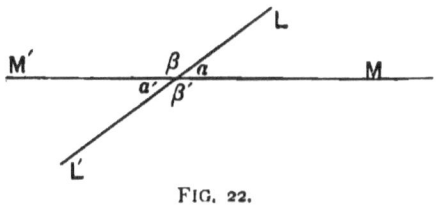

FIG. 22.

Two angles whose sum is a straight angle are called **supplemental**; two angles whose sum is a round angle we may call **explemental**. Two angles as α and α', the arms of the one being counter to the arms of the other, are called *opposite*, or *vertical*, or counter.

Theorem IX. — *When two rays meet, the opposite angles formed are equal.*

For $\alpha + \beta = S$ (a straight angle) (why?); and $\alpha' + \beta = S$ (why?).

Hence $\alpha + \beta = \alpha' + \beta$ (why?); therefore $\alpha = \alpha'$. Similarly let the student show that $\beta = \beta'$.   Q. E. D.

An important special case is when the *adjacents*, α and β, are *equal*. Each then is *half* of a straight angle, and therefore *one fourth* of a round angle; and each is called a **right angle**. Now let the student show that if $\alpha = \beta$, then $\alpha' = \beta$ and $\alpha = \beta'$, or

Corollary. When two intersecting rays make two equal adjacent angles, they make all four of the angles equal (Fig. 23).

Def. Rays that make right angles with one another are called **normal** (or perpendicular) to each other. N.B. The normal relation is *mutual*. How?

Def. Two angles whose sum is a right angle are called complemental.

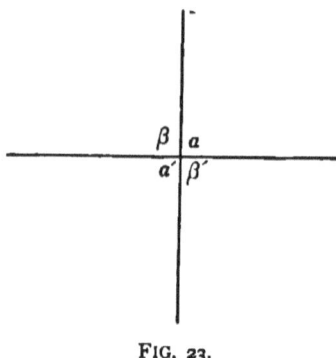

FIG. 23.

42. Are we sure that through any point on a ray we can draw a normal to the ray? Let O be any point on the ray LL' (Fig 24). Let any half-ray, pivoted at O, start

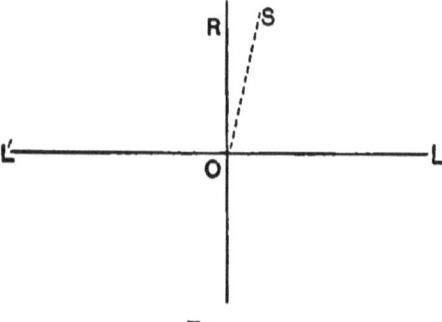

FIG. 24.

from the position OL and turn counter-clockwise into the position OL'. At first the angle on the right is *less* than the angle on the left, at last it is *greater;* the plane, the turning, and the angle are all continuous; hence in passing from the stage of being less to the stage of being greater, it passes through the stage of equality. Let OR be its position in this stage; then $\angle LOR = \angle ROL'$; *i.e.* OR is normal to LL'. Moreover, in no other position, as OS, is the ray normal to LL'; for LOS is not $= LOR$ unless OS falls on OR, but is less than LOR when OS falls within the angle LOR, while SOL' is greater than LOR; hence LOS and SOL' are not equal; *i.e.* OS is not normal to LL' when OS falls not on OR. Similarly, when LOS is greater than LOR. Hence

Theorem X. —*Through a point on a ray one, and only one, ray can be drawn normal to the ray.*

43. *Def.* A ray through the vertex of an angle, and forming equal angles with the arms of the angle, is called the *inner* Bisector or **mid-ray** of the angle. The inner bisector of an adjacent supplemental angle is called the *outer* bisector of the angle itself. Thus OI bisects *innerly* and OE bisects *outerly* the angle AOB (Fig. 25).

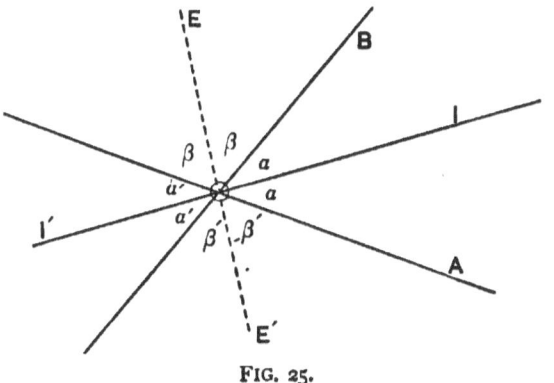

FIG. 25.

Exercise. Prove that there is one and only one such inner mid-ray.

44. Theorem XI. — *The inner Bisector of an angle bisects also its explement innerly.*

Proof. Let OI bisect $\angle AOB$ innerly; then $\angle AOI = \angle IOB$; call each α; then $\alpha + BOI' = \alpha + AOI'$ (why?); take away α; then $BOI' = AOI'$ (why?); *i.e.* the ray II' bisects innerly the angle BOA, the explement of AOB. Show that the angles marked α' are equal.

45. Theorem XII. — *The inner and outer Bisectors of an angle are normal to each other.*

Proof. Let OI and OE bisect (Fig. 25) innerly and outerly the angle AOB. Then, by definition, the angles marked α are equal, and the angles marked β are equal; also the sum of $+ \alpha + \alpha + \beta + \beta = S$; hence $\alpha + \beta = \tfrac{1}{2}S$; or, $IOE =$ a right angle. Q. E. D.

TRIANGLES.

46. Thus far we have treated only of rays intersecting in a single point. But, in general, three rays L, M, N (Fig.

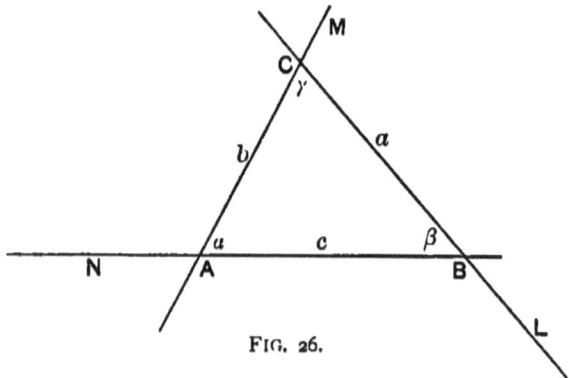

Fig. 26.

26) will meet in three points, since each pair will meet in one point, and there are three pairs: (MN), (NL), (LM).

Denote these points by A, B, C. Then the figure formed by these three rays is called a **triangle**, trigon, or three-side. A, B, C are its **vertices**; α, β, γ its *inner* **angles**; BC, CA, AB, its *inner* **sides**, or simply its **sides**. Its angles and sides are called its *parts*. It is the simplest closed rectilinear figure, and most important. If instead of taking three rays we take three points A, B, C, then we may join them in pairs by rays; and since there are three pairs, BC, CA, AB, then there are three rays, which we may name L, M, N. Thus we see that three points determine three rays, just as three rays determine three points. This equivalent determination of the figure by the same number of points as of rays makes the figure unique and especially important. We denote it by the symbol $\triangle$. We now ask, When are two triangles congruent?

47. Theorem XIII. — *Two $\triangle$ having two sides and the included angle of the one equal respectively to two sides and the included angle of the other are congruent.*

The **data** are: Two $\triangle$, ABC and $A'B'C'$, having the three equalities, $AB = A'B'$, $AC = A'C'$, $\alpha = \alpha'$ (Fig. 27).

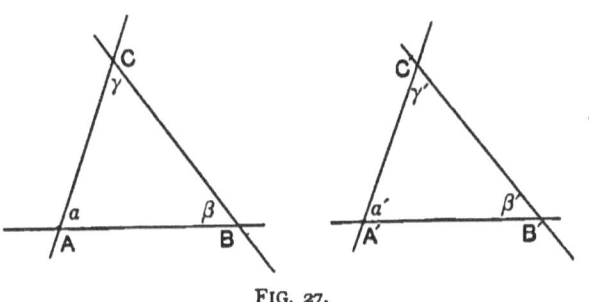

FIG. 27.

Proof. Fit the angle α on the angle α'; this is possible, because the angles are equal and congruent. Then A falls on A'; also the point B falls on B' (why? Because $AB = A'B'$), and C falls on C' (why?). Hence the three vertices of the two △ coincide in pairs; therefore the three sides of the two △ coincide in pairs (why? Because through two points, as $A(A')$ and $B(B')$, only one ray can pass). Q. E. D.

Corollary 1. The other parts of the two △ are equal or congruent in pairs of correspondents: $\beta = \beta'$, $\gamma = \gamma'$, $BC = B'C'$.

Corollary 2. Pairs of equal parts lie opposite to pairs of equal parts.

48. **Theorem XIV.** — *Two △ having two angles and the included side of the one equal respectively to two angles and the included side of the other are congruent* (Fig. 28).

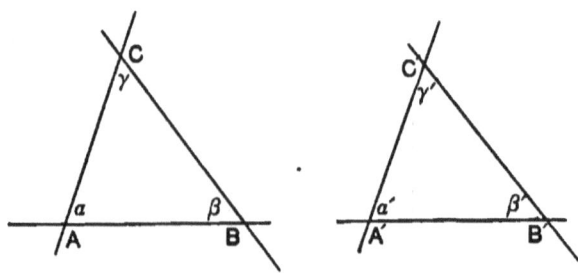

FIG. 28.

Data: Two △ ABC, $A'B'C'$, having $\alpha = \alpha'$, $\beta = \beta'$, $AB = A'B'$.

Proof. Fit AB on $A'B'$; this is possible (why?). Then α will fit on α' (why?), and β on β' (why?); *i.e.* the ray AC will fit on $A'C'$, and the ray BC on $B'C'$. Then the point C will fall on C' (why? Because two rays meet in only one point); *i.e.* the two △ fit exactly. Q. E. D.

TH. XVI.]   TRIANGLES.   37

49. We may now use the *conditions of congruence* thus far established to generate new notions that may be used in establishing other Theorems.

Def. The ray normal to a tract at its mid-point is called the **mid-normal** of the tract.

Theorem XV. — *Any point on the mid-normal of a tract is equidistant from its ends* (Fig. 29).

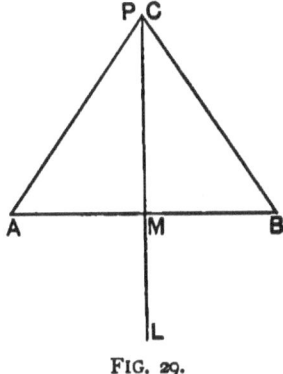

FIG. 29.

Data: AB a tract, M its mid-point, L the mid-normal, P any point on it.

Proof. Compare the △ APM and BPM. We have $AM = BM$ (why?). $PM = PM$, $\measuredangle AMP = \measuredangle BMP$ (why?); hence the △ are congruent (why?); and $PA = PB$. Q. E. D.

Def. A △ with two equal sides, like APB, is called **isosceles**; the third side is called the *base*, and its opposite angle the *vertical* angle.

50. **Theorem XVI.** — *The angles at the base of an isosceles △ are equal; and conversely.*

Data: ABC an isosceles △, AB its base, AC and BC its equal sides (Fig. 30).

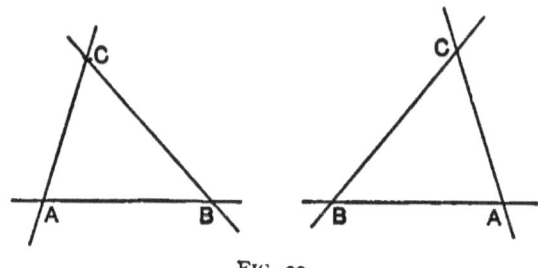

FIG. 30.

Proof. Take up the △ ABC, turn it over, and replace it in the position BCA. Then the two △ ACB and BCA have the equal vertical angles, C and C, also the side $AC = BC$ (why?) and $BC = AC$ (why?); hence they are congruent (why?), and the ∡ A = ∡ B. Q. E. D.

Conversely, *A △ whose basal angles are equal is isosceles.* Let the student conduct a proof quite similar to the foregoing.

Def. The ray through a vertex and the mid-point of the opposite side is called the **medial** of that side.

Corollary 1. In an isosceles △ the medial of the base is normal to it, and is the mid-ray of the vertical angle.

Corollary 2. When the medial of a side of a △ is normal to the side, the △ is isosceles. Prove it.

Corollary 3. When the medial of a side bisects the opposite angle, the △ is isosceles. Can you prove it?

LOGICAL DIGRESSION.

51. When the subject and predicate of a proposition are merely exchanged, the proposition is said to be *converted*, and the new proposition is called the *converse*. Thus X is Y; *conversely*, Y is X. In general, converses of true propositions are not true, but false. Thus, *The horse is an animal* is always correct, but *The animal is a horse* is generally false. A proposition remains true after simple conversion only when subject and predicate are properly quantified, thus: *All horses are some animals;* conversely, *Some animals are all horses*. Both propositions are correct and mean the same thing. But they are awkward in expression, and such forms are rarely or never used. When the quantifying word is *all* or its equivalent, the term is said to be taken *universally;* when it is *some* or its equivalent, the term is said to be taken *particularly*. Thus in the foregoing example horse is taken universally, but animal particularly. The only useful conversions are of propositions in which *both* subject and predicate are *universal*. In the great body of propositions only the subject is quantified universally, the quantifier is omitted from the predicate, but a particular one is understood. To show that a universal quantifier is admissible requires in general a distinct proof.

52. In order to convert an hypothetic proposition, we *exchange* hypothesis and conclusion. Thus, if X is Y, U is V; the converse is, if U is V, X is Y. All such hypothetic propositions may be stated *categorically*, thus: All cases of X being Y are cases of U being V; conversely, All cases of U being V are cases of X being Y. This converse is plainly false except when the quantifier *all* is admissible in the first predicate.

53. But while the converse of a true hypothetic proposition is generally false, the **contrapositive** is always true. This latter is formed by *exchanging* hypothesis and conclusion and *denying both*. Thus: If X is Y, then U is V; *contrapositive*, If U is not V, then X is not Y. Or, if a point is on the mid-normal of a tract, then it is equidistant from the ends of the tract; contrapositive, If a point is not equidistant from the ends of a tract, then it is not on the mid-normal of the tract.

---

54. Theorem XVII. — *An outer angle of a △ is greater than either inner non-adjacent angle.*

Data: Let ABC be any △, α' an outer angle, β' a non-adjacent inner one (Fig. 31).

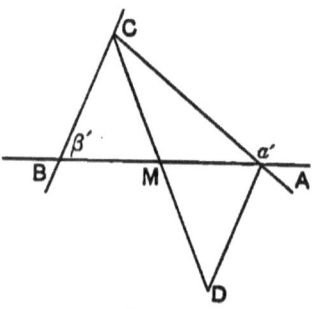

FIG. 31.

Proof. Draw the medial CM and lay off $MD = MC$; also draw AD. Then in the △ AMD and BMC we have $AM = BM$ (why?), $MD = MC$ (why?), and $\measuredangle AMD = \measuredangle BMC$ (why?); hence the △ are congruent (why?), and $\measuredangle MBC = \measuredangle MAD$ (why?). But $\measuredangle MAD$ is only part of the $\measuredangle \alpha'$; hence $\alpha' > \measuredangle MAD$ (why?); *i.e.* $\alpha' > \beta'$. Q. E. D.

Similarly, prove that $\alpha' > \gamma$.

55. Theorem XVIII. — *If two sides of a △ are unequal, then the opposite angles are unequal in the same sense (i.e. the greater angle opposite the greater side)* (Fig. 32).

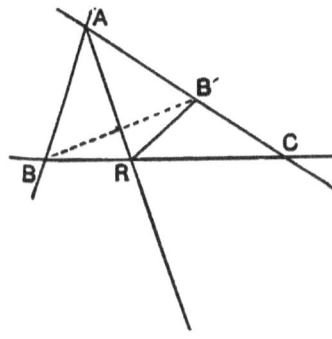

FIG. 32.

Data: ABC a △, $AC > AB$, AR the mid-ray of the angle at A, AB' laid off $= AB$.

Proof. ABR and $AB'R$ are congruent (why?); hence $\measuredangle ABR = \measuredangle AB'R$ (why?); but $\measuredangle AB'R > C$ (why?); i.e. $\measuredangle ABC > \measuredangle ACB$. Q. E. D.

Conversely, *If two angles of a △ are unequal, the opposite sides are unequal in the same sense.*

Proof. The opposite sides are not equal; for when the sides are equal, the opposite angles are equal (Theorem XVI.), and *contrapositively*, when the angles are unequal, the opposite sides are unequal. Then, by the preceding Theorem, the greater angle lies opposite the greater side.

56. Join BB'; then AR is the mid-normal of BB' (why?), and hence angle $CBB' =$ angle $BB'R$ (why?). Hence angle $BB'C > B'BC$ (why?); hence $BC > B'C$ (why?). But $B'C = AC - AB$; hence $BC > AC - AB$; i.e.

Theorem XIX. — *Any side of a △ is greater than the difference of the other two.*

Add AB to both sides of this inequality and there results $AB + BC > AC$; *i.e.*

Theorem XX. — *Any side of a △ is less than the sum of the other two.*

This fundamental Theorem is here proved on the supposition that $AB < AC$; if AB were $= AC$ or $> AC$, it would need no formal proof.

57. Theorem XXI. — *A point not on the mid-normal of a tract is not equidistant from the ends of the tract.*

Data: AB the tract, MN the mid-normal, Q any point not on MN (Fig. 33).

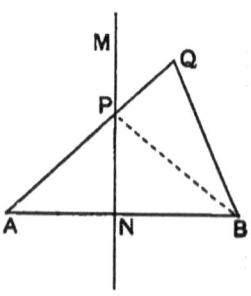

FIG. 33.

Proof. Draw QA and QB; one of them, as QA, must cut MN at some point, as P. Then $QB < QP + PB$ (why?), and $PB = PA$ (why?); hence $QB < QP + PA$; *i.e.* $QB < QA$. Q. E. D.

Of what Theorem is this the converse?

If now we seek for a point equidistant from A and B, we can find it on the mid-normal of AB and *only* there; hence *the locus of a point equidistant from the ends of a tract is the mid-normal of the tract.*

58. Theorem XXII.—*Two △ with the three sides of the one equal respectively to the three sides of the other are congruent.*

Data: ABC and $A'B'C'$ the two △, and $AB = A'B'$, $BC = B'C'$, $CA = C'A'$ (Fig. 34).

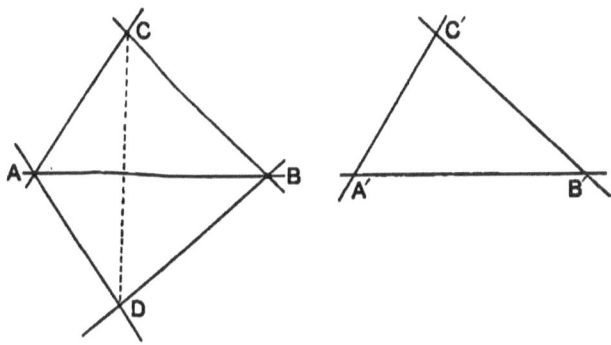

FIG. 34.

Proof. Turn the △ $A'B'C'$ over and fit $A'B'$ on AB so that C' shall fall (say) at D. Draw CD. Then A and B are on the mid-normal of CD (why?); hence the ray AB is the mid-normal of CD (why?); hence the angle $CAB =$ angle DAB, and angle $CBA =$ angle DBA (why?). Hence the △ are congruent (why?). Q. E. D.

N.B. As to when the △ must be turned round and when turned over, see Art. 94.

59. Theorem XXIII.—A. *From any point outside of a ray one normal may be drawn to the ray.*

Data: P the point, LL' the ray (Fig. 35).

Proof. From P draw a ray far to the left, as PA, making the angle $PAL >$ angle PAL'. Now let the ray turn about P as a pivot into some position far to the right, making angle $PA'L <$ $PA'L'$. The plane, the angle, the motion, all being continuous, in passing from the stage of being unequal in one sense to the stage of being unequal in the opposite sense, the angles made by the moving ray with the fixed ray must have passed through the stage of equality. Let PN be

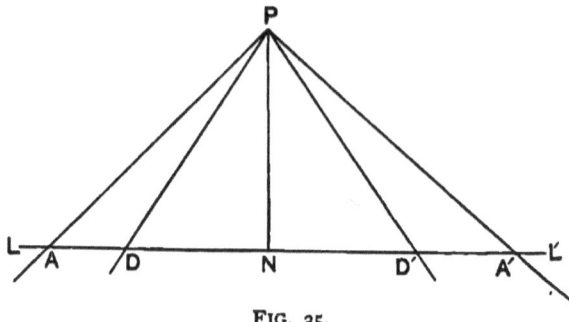

FIG. 35.

the ray in this position so that angle $PNL =$ angle PNL'; then each is a right angle by Definition, and PN is normal to LL'. Q. E. D.

B. *There is only one ray through a fixed point and normal to a fixed ray.*

Proof. Any other ray than PN, as PD, is not normal to LL'; for the outer angle PDL is $>$ the right angle PND (why?). Q. E. D.

C. *The normal tract PN is shorter than any other tract from P to the ray LL'.*

Proof. For the right angle at N is $>$ angle PDN (why?); hence $PN < PD$ (why?). Q. E. D.

D, E. *Equal tracts from point to ray meet the ray at equal distances from the foot of the normal; and conversely.*

Proof. For, if DPD' be isosceles, then the normal PN is the medial of the base (why?).

F. *Two, and only two, tracts of given length can be drawn from a point to a ray.*

Proof. For two, and only two, points are on the ray at a given distance from the foot of the normal.

G. *Of tracts drawn to points unequally distant from the foot of the normal, the one drawn to the remotest is the longest.*

Proof. In the △ PDA, angle $PDA > PAD$ (why?); hence $PA > PD$ (why?). Q. E. D.

Similarly, $PA' > PD$.

H. *Equal tracts from the point to the ray make equal angles with the normal from the point to the ray and also equal angles with the ray itself;* and conversely.

I. *Of unequal tracts from the point to the ray, the longest makes the greatest angle with the normal and the least with the ray.*

Let the student conduct the proof of H and I.

60. **Theorem XXIV.** — *Two △ having two angles and an opposite side of one equal respectively to two angles and an opposite side of the other are congruent.*

Data: ABC and $A'B'C'$ two △ having $AB = A'B'$, angle α = angle α', angle γ = angle γ' (Fig. 36).

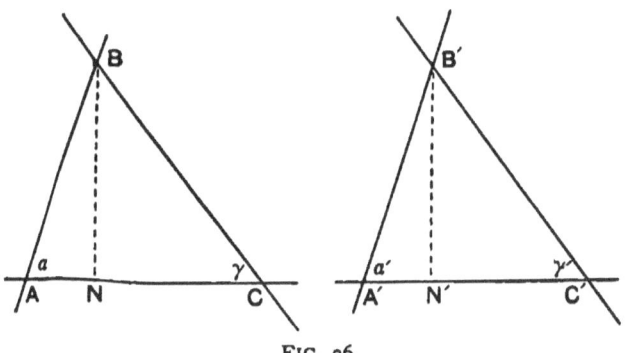

Fig. 36.

46   GEOMETRY.   [TH. XXIV.

Proof. Fit α' on α; then B' falls on B (why?), and $A'C'$ falls along AC. Draw the normal BN. Then BC and BC' make the same angle, $\gamma = \gamma'$, with the ray AN; hence they are $=$ and meet the ray in the same point (why?); *i.e.* C' falls on C; *i.e.* the △ are congruent.   Q. E. D.

61. We now come to the so-called **ambiguous** case, of two △ with two sides and an opposite angle in one equal to the two sides and the corresponding opposite angle in the

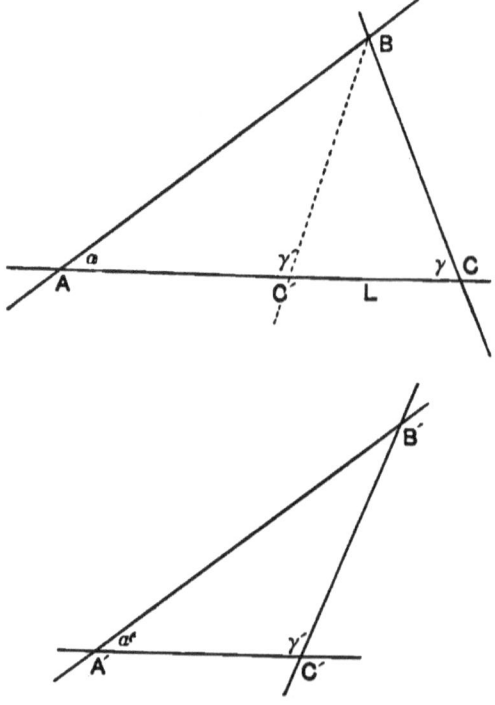

FIG. 37.

other. Let ABC and $A'B'C'$ (Fig. 37) be the two △, with $AB = A'B'$, $BC = B'C'$, and angle $\alpha =$ angle α'. Fit α' on

TRIANGLES.

a; then $A'B'$ falls on AB, B' on B; but since from a point B (B') we may draw two equal tracts to the ray AL, the side $B'C'$ may be either of these equals and *may or may not* fall on BC. In general, then, we cannot prove congruence in this case. But if BC be $> AB$, then angle $a >$ angle γ (why?), and there is *only one* tract on the right of AB drawn from B to the ray AC and equal to BC; the *other* tract equal to BC must be drawn outside of AB and to the left. Hence in this case, when the angle lies opposite the greater side, the △ are congruent. Hence

Theorem XXV. — *Two △ having two sides and an angle opposite the greater in one equal to two sides and an angle opposite the greater side in the other are congruent.*

Corollary. Two right △ having a side and any other part of one equal to a side and the corresponding part of the other are congruent.

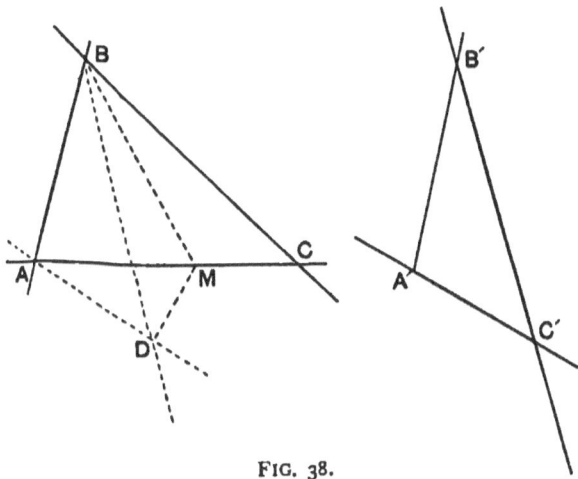

FIG. 38.

62. We have seen (Art. 47) that when two △ have two sides and included angle in one equal to two sides and

included angle in the other, they are congruent. But what if the included angles are not equal? Let ABC and $A'B'C'$ be the two △, having $AB = A'B'$, $BC = B'C'$, but $\beta > \beta'$. Slip the upper film of the plane along until $A'B'$ fits on AB and let C' fall on D. Draw the mid-ray BM of the angle CBD, let it cut AC at M, and draw DM. Then the △ CBM and DBM are congruent (why?); hence $AM + MD = AC$ (why?), and $AC > AD$, or $AC > A'C'$. Hence

Theorem XXVI. — *Two △ having two sides of one equal to two sides of the other, but the included angles unequal, have also the third sides unequal, the greater side lying opposite the greater angle.*

Conversely, *Two △ having two sides in one equal to two sides in the other, but the third sides unequal, have the included angles also unequal, the greater angle being opposite the greater side.*

Proof. The included angles are *not equal;* for if they were equal, the △ would be congruent (why?) and the three sides would be equal. Hence the included angles are unequal, and the relation just established holds; namely, the greater angle lies opposite a greater side. Q. E. D.

63. **Theorem XXVII.** — *Every point on a mid-ray of an angle is equidistant from its sides.*

Data : O the angle, MM' the mid-ray, P any point on it.

Proof. From P draw the normals PC and PD; they are (Fig. 39) the distances of P from the ends of the angle. Then the △ POC and POD are congruent (why?); hence $PD = PC$. Q. E. D.

Th. XXVIII.]   TRIANGLES.   49

Conversely, *A point equidistant from the ends of an angle is on a mid-ray of the angle* (Fig. 39).

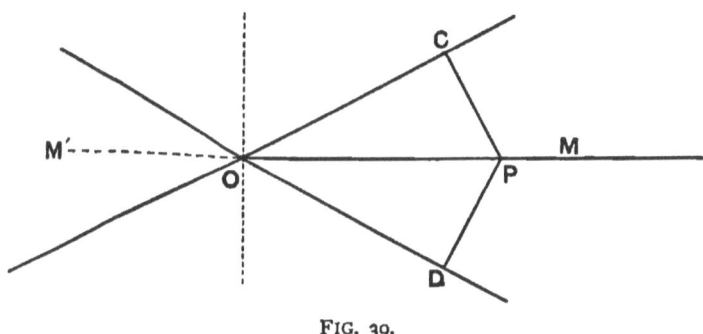

FIG. 39.

Proof. If $PC = PD$, then the △ POC and POD are congruent (why?) ; hence angle $POD =$ angle POC. Q. E. D.

Accordingly we say that the *mid-rays of an angle are the locus of a point equidistant from its ends*.

*64. It is just at this stage in the development of the doctrine of the Triangle that we are compelled to halt and introduce a new concept before we can proceed any further. The necessity of this step will appear from what follows (which may, however, be omitted on first reading, at the option of teacher or student).

Def. Two △ not congruent are called **equivalent** when they may be cut up into parts that are congruent in pairs.

Theorem XXVIII. — *Any △ is equivalent to another △ having the sum of two of its angles equal to the smallest angle of the given △.*

Data : ABC the △, a the least angle (Fig. 40).

Proof. Through M, the mid-point of BC, draw AM and make $MD = MA$. Then the $\triangle$ ACM and DBM are congruent (why?), the part AMB is common to ABC and ABD, and the sum of the angles ADB and BAD = angle BAC. Q. E. D.

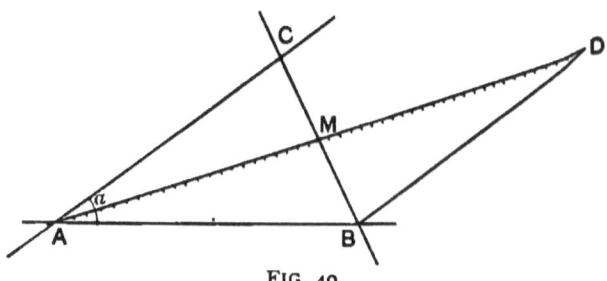

FIG. 40.

Corollary. The sum of the angles in the new $\triangle$ is equal to the sum of the angles in the old $\triangle$.

*65. We may now repeat this process, applying it to the smallest angle, as A, of the $\triangle$ ABD. In the new $\triangle$ ABE the smallest angle, as A, cannot be greater than $\frac{1}{4}$ of the original angle α in ABC; after n repetitions of this process we obtain a $\triangle$, as ALB, in which the sum of the angles A and L cannot be $> \frac{1}{2^n}$ of the original angle α in the $\triangle$ ABC. By making n as large as we please, we make $\frac{1}{2^n}$ as small as we please, and so we make $\frac{1}{2^n}$ of angle α smaller than any assigned magnitude no matter how small. Meantime the other angle B has indeed grown larger and larger, but has remained $<$ a straight angle. Hence the sum of the angles in the $\triangle$ ALB cannot exceed a straight angle by any amount however small; but the sum of the angles in ALB = sum of the angles in ABC; hence

Theorem XXIX. — *The sum of the angles in any $\triangle$ cannot exceed a straight angle by any finite amount.*

Corollary 1. The outer angle of a △ is *not less* than the sum of the inner non-adjacent angles.

Corollary 2. From any point outside of a ray there may be drawn a ray making with the given ray an angle small at will.

Proof. From P draw any ray PA, and lay off $AB = PA$ (Fig. 41). Then the angle PBA is not greater than

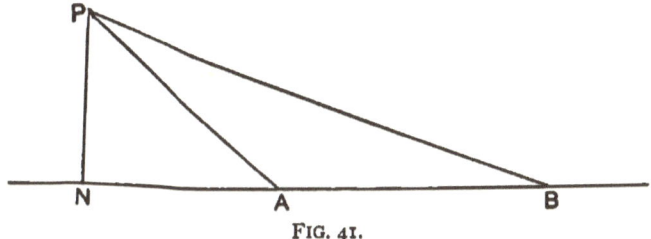

FIG. 41.

$\frac{1}{2} PAN$ (why?); now lay off $BC = PB$ (why?); then angle PCB is not $> \frac{1}{2}$ angle PBA (why?); proceeding this way, we obtain after n constructions an angle PLN not $> \frac{1}{2^n}$ of the angle PAN, and by making n large enough we may make this $\frac{1}{2^n}$ as small as we please. Q. E. D.

*66. **Theorem XXX.** — *If the sum of the angles in any △ equals a straight angle, then it equals a straight angle in every △* (Fig. 42).

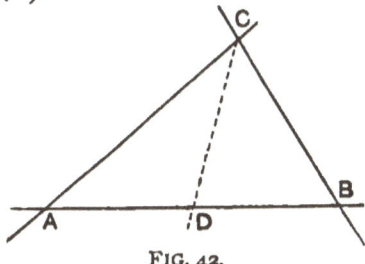

FIG. 42.

Hypothesis: ABC a $\triangle$ with the sum of its angles $A + B + C = S$.

Proof. (1) Draw any ray through C, as CD. Then if the sum of the angles in the $\triangle$ ACD and BCD be $S-x$ and $S-y$, x and y being any definite magnitudes however small, then on adding these sums we get $2S-(x+y)$; and on subtracting the sum, S, of the supplemental angles at D we get $S-(x+y)$ for the sum of the angles of the $\triangle$ ABC. Now if this sum be S, then x and y must each be O; *i.e.* the sum of the angles in each of the $\triangle$ ACD and BCD is S. Now draw DE and DF; in each of the four small $\triangle$ the sum of the angles is still $= S$. (2) We may now make a $\triangle$ as large as we please and of any shape whatever, but the sum of the angles will remain $= S$. For, take the same $\triangle$ ABC, and draw CD normal to AB. Then the sum of the (Fig. 43) angles in the $\triangle$ ACD is S, as has

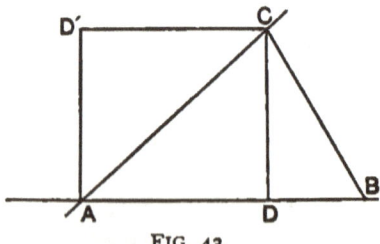

FIG. 43.

been shown above; also angle D is a right angle; hence the angles A and ACD are complementary. Now along AC fit another $\triangle$ ACD' congruent with ACD; then all the angles of the quadrilateral $ADCD'$ are right, and the figure is called a **rectangle**. Now we can place horizontally side by side as many of these rectangles, all congruent, as we please, say p of them; we can also place as many of them vertically, one upon another, as we please, say q of them;

and we can then fill up the whole figure into a new rectangle, as large as we please. About each inner junction-point of the sides of the rectangles there will be four right angles plainly. Now connect the two opposite vertices, as A and Z, of this rectangle. So we get two congruent right $\triangle$, in each of which the sum of the angles is S. Then any $\triangle$ that we cut off from this right $\triangle$ will, by the foregoing, have the sum of its angles equal to S. Since p and q are entirely in our power, we may make in this way any desired right $\triangle$ and from it cut off any desired oblique $\triangle$, with the sum of its angles $= S$. Q. E. D.

Hence *either no $\triangle$ has the sum of its angles $= S$*, or **every $\triangle$ has the sum of its angles $= S$**.

67. A logical choice between these alternatives is impossible, but the matter may be cleared up by the following considerations:

Across any ray LM draw a transversal T, cutting LM at O, and making the angles α, β, γ, δ. Through any point, as O', of T draw a ray (Fig. 44) $L'M'$ making angle $\alpha' = \alpha$.

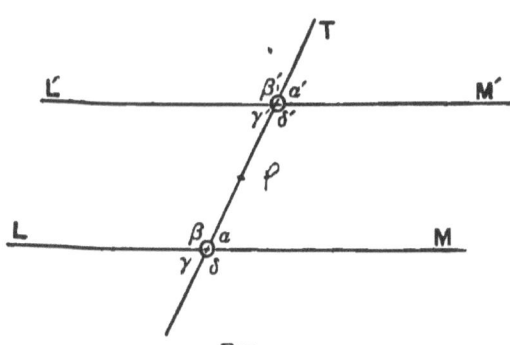

FIG. 44.

This is evidently possible (why?). Then plainly $\beta' = \beta$, $\gamma' = \gamma$, $\delta' = \delta$, $\alpha' = \alpha$; they are called **corresponding** angles;

also α and γ', β and δ' are equal, — they are called **alternate** angles; also α and δ', as well as β and γ', are supplemental, — they are called **interadjacent** angles.

68. Now let P be the mid-point of OO'; on it as a pivot turn the whole right side of the plane round through a straight angle until O falls on O', and O' falls on O. Then, since the angles about O and O' are equal as above stated, the half-ray OL will fall and fit on the half-ray $O'M'$, and the half-ray $O'L'$ on the half-ray OM. Accordingly, *if the rays LM and $L'M'$ meet on one side of the transversal T, they also meet on the other side of T.*

69. Three possibilities here lie open:

(1) The rays LM and $L'M'$ may meet on the left and also on the right of T, in *different* points.

(2) They may meet on the left and also on the right of T, in the *same* point.

(3) They may **not meet at all**.

No logical choice among these three is possible. But *in all regions accessible to our experience* the rays neither converge nor show any tendency to converge. Hence we *assume* as an

Axiom A. *Two rays that make with any third ray a pair of corresponding angles equal, or a pair of alternate angles equal, or a pair of interadjacent angles supplemental, are* **non-intersectors.**

70. But another query now arises. Is it possible to draw another ray through O' so close to L' that it will not meet OL however far both may be produced? Here again it is impossible to answer from pure logic. An appeal to experience is all that is left us. This latter testifies that no ray can be drawn through O' so close to $O'L'$ as not to approach and finally meet the ray OL. Hence we *assume* as another

Axiom B. *Through any point in a plane* **only one** **non-intersector** *can be drawn for a given straight line.*

This single non-intersector is commonly called the **parallel**, through the point, to the straight line.

71. It cannot be too firmly insisted, nor too distinctly understood, that the existence of any non-intersector at all, and the existence of only one for any given point and given ray, are both assumptions, which cannot be proved to be facts. The best that can be said of them, and that is quite good enough, is that they and all their logical consequences accord completely and perfectly with all our experience as far as our experience has hitherto gone. Even then, if there be any error in our assumptions, we have thus far been utterly unable to find it out.

A geometry that should reject either or both of these assumptions would have just as much logical right to be as the geometry that accepts them, and such geometries lack neither interest nor importance. They may be called **Hyper-Euclidean** in contradistinction from this of ours, which from this point on is Euclidean (so-called from the Greek master, Euclides, who distinctly enunciated the *equivalent* of our Axioms in a Definition and a Postulate).

NOTE. — Observe the relation of Axioms **A** and **B**: *the one is the converse of the other.*

Observe also that the necessity of assuming the first lies in our ignorance of the *indefinitely great*, and the occasion of assuming the other lies in our ignorance of the *indefinitely small*. See Note, Art. 301.

72. Accepting our Axioms as at least *exacter than any experiment we can make*, we may now easily settle the ques-

tion as to the sum of the angles in a △. Let ABC be any △; through the vertex C draw the one parallel to the base AB. Then $\alpha = \alpha'$, $\beta = \beta'$ (why?); also $\alpha' + \gamma + \beta' = S$; hence $\alpha + \gamma + \beta = S$; i.e. (Fig. 45)

Theorem XXXI. — *The sum of the angles in a △ is a straight angle.*

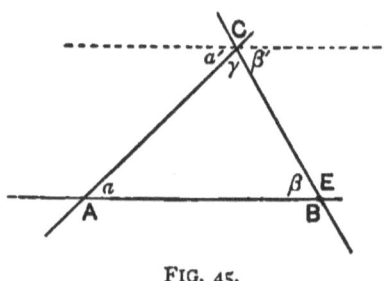

FIG. 45.

Corollary 1. The outer angle E equals the sum of the inner non-adjacent angles α and γ (why?).

Corollary 2. If two angles of a △ be known, the third is also known.

Corollary 3. If two △ have two angles, or the sum of two angles of the one equal to two angles, or the sum of two angles of the other, then the third angles are equal.

Corollary 4. To know the three angles of a △ is not to know the △ completely, for many △ may have the same three angles. Such △ are **similar**, as we shall see, but are not congruent; they are alike in **shape**, but not in **size**.

73. Next to normality, parallelism is the most important relation in which rays can stand to each other, and we must now use the new relation in the generation of new concepts.

Theorem XXXII. — *Parallel Intercepts between parallels are equal.*

Data: L and L', M and M', two pairs of parallels (Fig. 46).

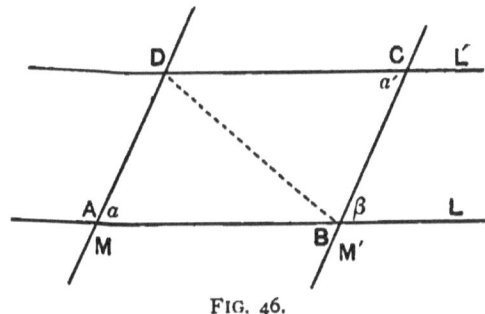

FIG. 46.

Proof. Draw BD. Then the △ ABD and CDB are congruent (why?), and $AB = CD$, $BC = DA$. Q. E. D.

Def. The figure $ABCD$ formed by two pairs of parallel sides is called a **parallelogram**, and may be denoted by the symbol ▱.

A join of opposite vertices, as BD, is called a **diagonal**.

74. **Theorem XXXIII.** — **Properties of the parallelogram.**

A. *The opposite sides of a parallelogram are equal.*

This has just been proved.

B. *The opposite angles of a parallelogram are equal.*

Proof. $\alpha = \beta$ (why?); $\beta = \alpha'$ (why?); hence $\alpha = \alpha'$. Q. E. D.

Corollary. Adjacent angles of a parallelogram are supplementary.

C. *Each diagonal of a parallelogram cuts it into two congruent* △. Prove it.

58   GEOMETRY.   [Th. XXXIV.

D. *The diagonals of a parallelogram bisect each other* (Fig. 47).

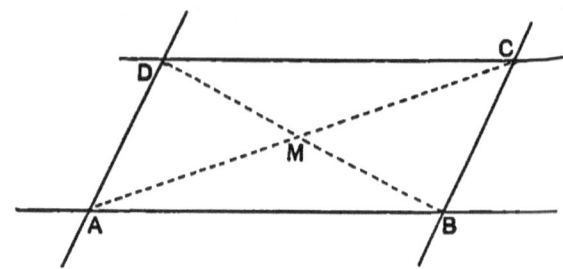

Fig. 47.

Proof. The △ *AMB* and *CMD* are congruent (why?); hence $AM = CM$, $BM = DM$. Q. E. D.

75. We may now convert all the foregoing propositions and obtain as many criteria of the parallelogram.

Theorem XXXIV. — A'. *A 4-side with its opposite sides equal is a parallelogram.*

Data: $AB = CD$, $AD = CB$ (Fig 48).

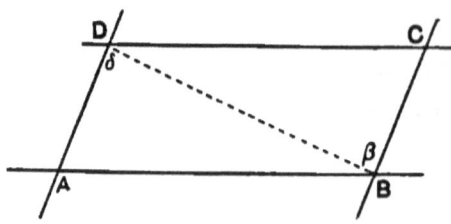

Fig. 48.

Proof. Draw *BD*. Then *ABD* and *CDB* are congruent (why?); hence $\beta = \delta$; and *AD* and *CB* are parallel; similarly, *AB* and *CD* are parallel; hence *ABCD* is a parallelogram. Q. E. D.

Th. XXXIV.] PARALLELOGRAMS. 59

B'. *A 4-side with opposite angles equal is a parallelogram.*

Data: $\alpha = \alpha'$, $\beta + \beta' = \gamma + \gamma'$ (Fig. 49).

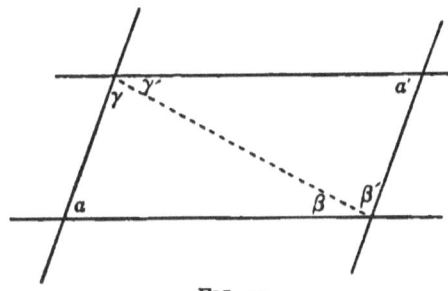

FIG. 49.

Proof. Since $\alpha = \alpha'$, $\beta + \gamma = \beta' + \gamma'$ (why?). Hence $\beta = \gamma'$, $\beta' = \gamma$; *i.e.* opposite sides are parallel, the 4-side is a parallelogram. Q. E. D.

C'. *A 4-side that is cut by each diagonal into two congruent △ is a parallelogram.*

For the opposite angles must be equal (why?); hence, etc. Q. E. D.

D'. *A 4-side whose diagonals bisect each other is a parallelogram.*

For the opposite sides are equal, being opposite equal angles in congruent △; hence, etc. Q. E. D.

E'. *A 4-side with one pair of sides equal and parallel is a parallelogram.*

For the other two sides are equal and parallel (why?); hence, etc. Q. E. D.

76. The foregoing properties and criteria of the parallelogram illustrate excellently the nature of a *definition*. This latter defines or *bounds off* by stating something that is true of the thing defined, but of nothing else. Accordingly, the characteristic of every definition or definitive property is that the proposition that states it may be converted simply. Thus:

Every parallelogram is a 4-side with opposite angles equal; and *conversely*, every 4-side with opposite angles equal is a parallelogram.

Not every property is definitive, and hence not every property may be used as test or criterion.

77. Special Parallelograms.

Def. An equilateral parallelogram is called a **rhombus**.

Theorem XXXV. — *The diagonals of a rhombus are normal to each other.*

Let the student conduct the proof suggested by the figure (Fig. 50).

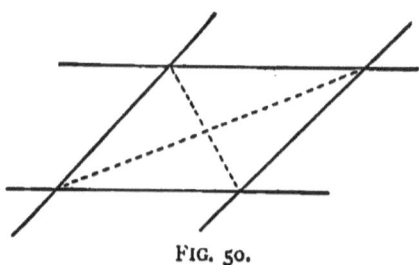

FIG. 50.

Conversely, *A parallelogram whose diagonals are normal to each other is equilateral, or a rhombus.* Let the student supply the proof.

78. *Def.* An equiangular parallelogram is called a **rectangle** (for all the angles are *right* angles).

Theorem XXXVI. — *The diagonals of a rectangle are equal* (Fig. 51).

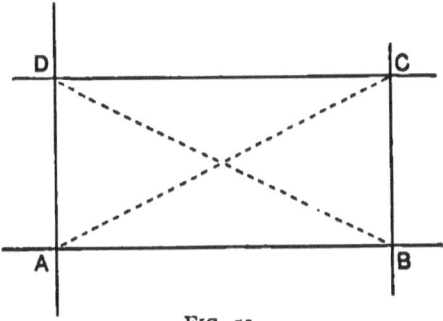

FIG. 51.

For the △ ABC and BAD are congruent (why?) ; hence $AC = BD$. Q. E. D.

Conversely, *A parallelogram with equal diagonals is equiangular, or a rectangle.*

For the △ ABC and BAD are again congruent, though for another reason. What reason ?

79. *Def.* A parallelogram both equilateral and equiangular is called a **square**.

Theorem XXXVII. — *The diagonals of a square are equal and normal to each other.*

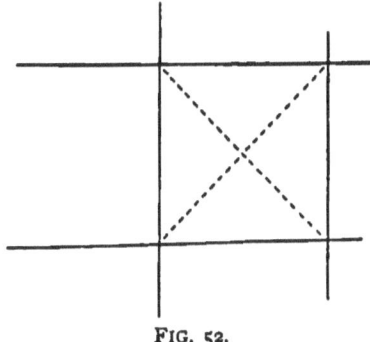

FIG. 52.

For the square, being both rhombus and rectangle, has all the definitive properties of both. Or the student may prove the proposition directly from the figure (Fig. 52), as well as its converse:

A parallelogram with diagonals equal and normal to each other is a square.

80. Can we convert Theorem XXXII. and prove that equal intercepts between parallels are parallel? Manifestly no (Fig. 53), for from the point C we may draw two equal

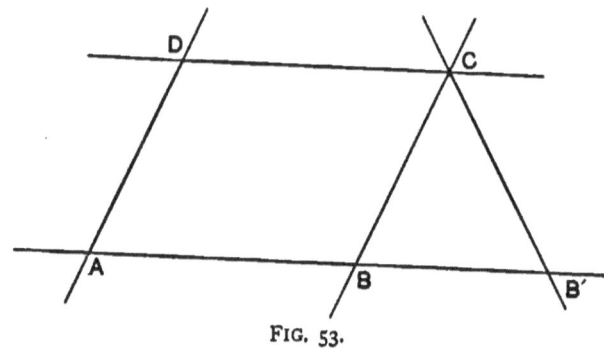

FIG. 53.

tracts to the other parallel, the one CB parallel to AD, the other CB' sloped at the same angle to the parallels but in opposite ways. We may call CB' *anti-parallel* to AD, and the figure $AB'CD$ an **anti-parallelogram**. Since from any point C only two equal tracts, or tracts of given length, may be drawn to the other parallel through A, we have the

Theorem XXXVIII. — *Equal intercepts between parallels are either parallel or anti-parallel.*

Corollary 1. Adjacent angles of an anti-parallelogram are alternately equal or supplemental.

Corollary 2. Anti-parallels prolonged meet at the vertex of an isosceles △.

THE GENERAL QUADRILATERAL OR 4-SIDE.

81. A Quadrilateral is determined by four intersecting rays. These determine six points, the four *inner* vertices, C, D, E, F, and the two *outer* ones, A, B. The cross-rays, CE, DF, AB, are the diagonals, CE and DF *inner*, AB *outer*. Commonly the outer diagonal is little used, and the inner ones are called *the* diagonals. When none of the angles C, D, E, F, of the 4-side is greater than a straight angle, the 4-side is called the **normal**, as $CDEF$. It is the only form ordinarily considered. The other two forms are (2) the *crossed*, $ACBE$, and (3) the *inverse*, $ADBF$ (Fig. 54). For all forms let the student prove

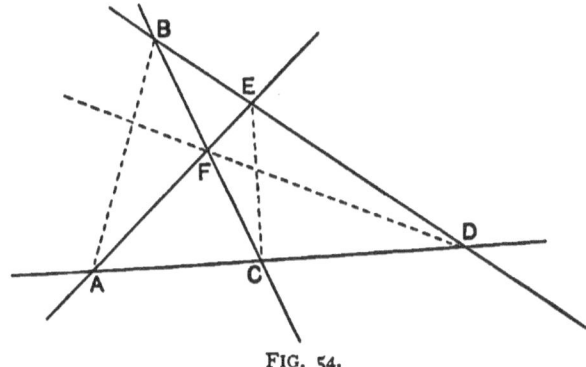

FIG. 54.

Theorem XXXIX. — *The sum of the inner angles of a 4-side is a round angle.*

Corollary. When two angles of a 4-side are supplemental, so are the other two.

82. Theorem XL. — *The angles between two rays equal the angles between two normals to the rays.*

Data: OL and OM any two rays, PA and PB any two normals to them (Fig. 55).

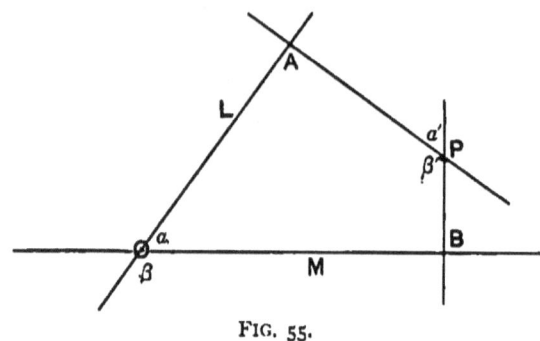

FIG. 55.

Proof. The angles at A and B are right angles and therefore supplemental (why?); hence $\alpha = \alpha'$, and $\beta = \beta'$. Q. E. D.

N.B. The 4-side with its *opposite angles supplemental* is very important and has received the name **encyclic** 4-side, for reasons to be seen later on (Arts. 126-7).

THREE OR MORE PARALLELS.

83. Theorem XLI. — *Three parallels that make equal intercepts on one transversal, make equal intercepts on any transversal.*

Data: L, M, N, three parallels, and $AB = BC$, and DEF any transversal (Fig. 56).

Proof. Draw $D'EF'$ parallel to ABC. Then $AB = BC$ (why?), $AB = D'E$ (why?), and $BC = EF'$ (why?); hence $D'E = EF'$ (why?), hence the △ DED' and FEF' are congruent (why?); hence $DE = EF$ (why?). Q. E. D.

84. Def. A 4-side formed by two parallels and two transversals is called a **trapezoid**. Thus $ACFD$ is a trapezoid. The parallel sides are called the **bases** (major and minor); the parallel through the mid-points of the transverse sides is the *mid-parallel*.

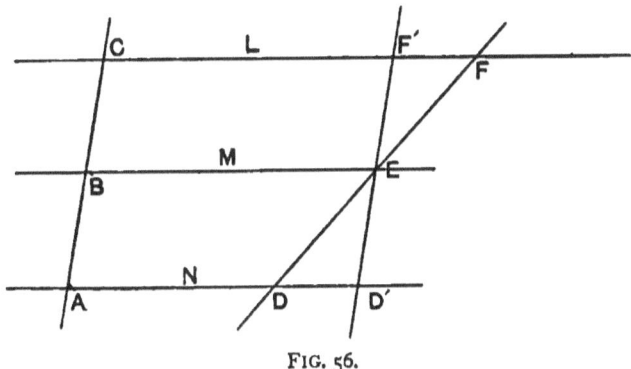

FIG. 56.

Theorem XLII. — *The mid-parallel of a trapezoid equals the half-sum of its bases.*

Let the student elicit the proof from the foregoing figure.

Corollary 1. A parallel to a base of a $\triangle$ bisecting one side bisects also the other. (*Hint.* Let D fall on A.)

Corollary 2. A ray bisecting two sides of a $\triangle$ is parallel to the third.

For only one ray can bisect two sides (why?), and we have just seen (Cor. 1) that a ray parallel to the base does this; hence, Q. E. D.

Corollary 3. The mid-parallel to the base of a $\triangle$ equals half the base.

85. Def. Three or more rays that pass through a point are said to **concur** or be concurrent.

Theorem XLIII. — *The medials of a △ concur.*

Data: ABC a △, AP and BQ two medials (Fig. 57).

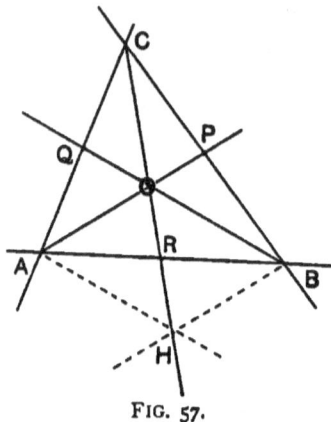

FIG. 57.

Proof. Draw a ray from C through O, the intersection of the two medials, and lay off $OH = CO$. Draw AH and BH; they are parallel to BQ and AP (why?); hence $AOBH$ is a parallelogram (why?); hence $AR = BR$ (why?). Hence COR is the third medial; *i.e.* the three medials pass through O. Q. E. D.

Corollary. Each medial cuts off a third from each of the other two. For $CO = 2OR$ (why?).

Def. The point of concurrence of the medials is called the **centroid** of the △. It is two-thirds the length of each medial from the corresponding vertex.

86. Theorem XLIV. — *The mid-normals of the sides of a △ concur.*

Data: ABC a △, L and M mid-normals to the sides BC and CA, meeting at S.

Proof. S is equidistant from B and C (why?), and from C and A (why?); hence S is equidistant from A and B (why?), or is on the mid-normal of AB (why?); hence the mid-normals concur (Fig. 58). Q. E. D.

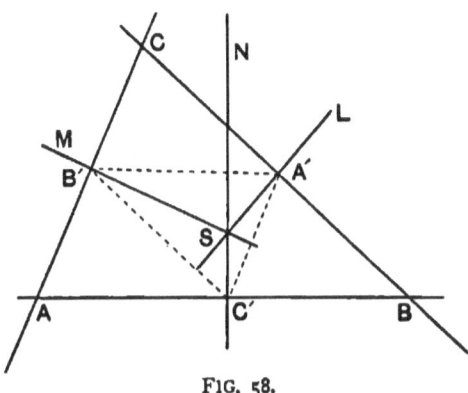

FIG. 58.

Corollary. S is equidistant from A, B, and C, and no other point in the plane is (why?).

Def. The point of concurrence of the mid-normals is called the **circumcentre** of the $\triangle$.

87. *Def.* A tract from a vertex of a $\triangle$ normal to the opposite side is called an **altitude** of the $\triangle$. Sometimes, when length is not considered, the whole ray is called the altitude.

Theorem XLV. — *The altitudes of a $\triangle$ concur.*

Proof. Using the preceding figure, draw the $\triangle$ $A'B'C'$. Its sides are parallel to the sides of ABC (why?); hence its altitudes are the mid-normals L, M, N; and these have just been found to concur. Also, since ABC may be any $\triangle$, $A'B'C'$ may be any $\triangle$; hence the altitudes of any $\triangle$ concur. Q. E. D.

Def. The point of concurrence of altitudes is called the **orthocentre** (or alticentre) of the △.

Def. In a right △ the side opposite the right angle is called the **hypotenuse** (= subtense = under-stretch).

Queries: Where do circumcentre and orthocentre lie: (1) in an acute-angled △? (2) in an obtuse-angled △? (3) in a right △?

88. Theorem XLVI.—*The inner mid-rays of the angles of a △ concur.*

Data: ABC a △, AL, BM, CN the inner mid-rays of its angles (Fig. 59).

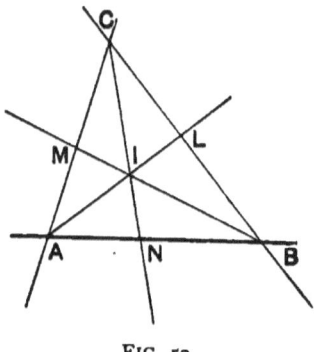

FIG. 59.

Proof. Let AL and BM intersect at I. Then I is equidistant from AB and AC, and from AB and BC (why?); hence I is equidistant from AC and BC; hence I is on the inner mid-ray of the angle C; *i.e.* the three inner mid-rays concur in I.   Q. E. D.

Def. The point of concurrence of the inner mid-rays is called the **in-centre** of the △.

[Th. XLVII.] EXERCISES I. 69

89. **Theorem XLVII.** — *The outer mid-rays of two angles and the inner mid-ray of the other angle of a $\triangle$ concur.*
Let the student conduct the proof (Fig. 60).

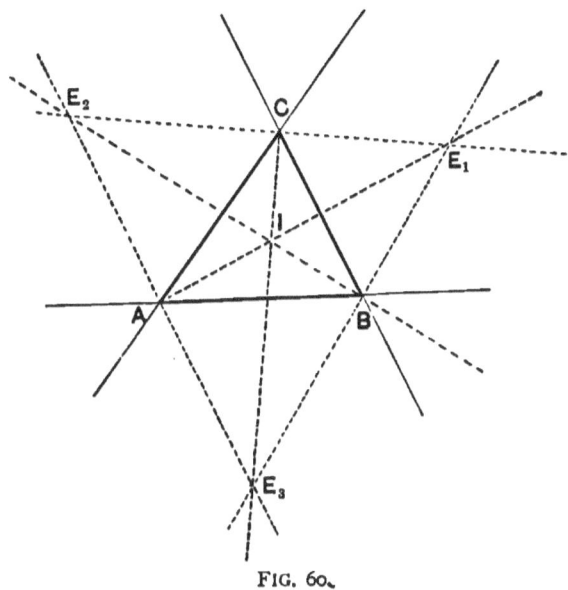

FIG. 60.

Def. The points of concurrence are called **ex-centres** of the $\triangle$: there are *three*.

EXERCISES I.

Little by little the student has been left to rely more and more upon his own resources of knowledge and ratiocination in the conduct of the foregoing investigations. He has now possessed himself of a large fund of concepts, and he must test his ability to wield, combine, and manipulate them in forging original proofs of theorems. Let him bear always in mind the fundamental logical principle that *every example of a general concept has all the marks of that general concept.* Let him begin his proof by stating precisely the *data*, the given or known facts, let him draw a corresponding diagram in order to have a clearer view of the spatial relations involved, let him note carefully what concepts are present in the proposition, let him draw auxiliary lines and introduce auxiliary concepts at pleasure. But let him exhaust simple means before trying more complicated, let him distinguish, by manner of drawing, the principal from the auxiliary rays, and especially let him be systematic and consistent in the literation of his figures.

1. How many degrees in a straight angle? In a right angle?

HISTORICAL NOTE. — For purposes of computation the round angle is divided into 360 equal parts called **degrees**, each degree into 60 equal **minutes** (partes *minutæ* primæ), each minute into 60 equal **seconds** (partes minutæ *secundæ*), denoted by °, ′, ″ respectively. This sexagesimal division is cumbrous and unscientific, but is apparently permanently established. It seems to have originated with the Babylonians, who fixed approximately the length of the year at 360 days, in which time the sun completed his circuit of the heavens. A *degree*, then, as is indicated by the name, which means *step* in Latin, Greek, Hebrew (*gradus*, βαθμος (or τμημα), ma'alah), was primarily the daily *step* of the sun eastward among the stars. The Chinese, on the other hand, determined the year much more exactly at 365¼ days, and accordingly, in defiance of all arithmetic sense, divided the circle into 365¼ degrees.

2. The angles of a △ are equal; how many degrees in each?

REMARK. — Such a △ is called *equiangular*, more commonly *equilateral*, but better still **regular**.

3. Show that this regular △ is equilateral.

4. One angle of a △ is a right-angle; the others are equal; how many degrees in each?

EXERCISES I.

5. One angle of a △ is twice and the other thrice the third; what are the angles?

6. Two angles of a △ are measured and found to be $46° 37' 24''$ and $52° 48' 39''$; what is the third?

7. One angle of a △ is measured to be $61° 22' 40''$; the others are computed to be $49° 34' 28''$ and $69° 2' 43''$; what do you infer?

8. A half-ray turns through two round angles counter-clockwise, then through half a right-angle clockwise, then through a straight angle counter-clockwise, then through $\frac{1}{8}$ of a round angle counter-clockwise, then through $\frac{7}{8}$ of a straight angle clockwise; what angle does it make in its final position with its original position?

9. O is a fixed point (called *origin*) on a ray, A and B are any pair of points, M their mid-point. Show and state in words that $2OM = OA + OB$.

10. A, B, C are three points on a ray, A', B', C' are mid-points of the tracts BC, CA, AB, and O is any point on the ray; show that $OA + OB + OC = OA' + OB' + OC'$.

11. A, B, C, D, O are points on a ray; A', B', C' are mid-points of AB, BC, CD; A'', B'', are mid-points of $A'B', B'C'$; M is the mid-point of $A''B''$; prove $8OM = OA + 3OB + 3OC + OD$.

12. What are the conditions of congruence in isosceles △? In right △?

13. In what △ does one angle equal the sum of the other two?

Def. A number of tracts joining consecutively any number of points (first with second, second with third, etc.) is called a *broken line*, or *train of tracts*, or **polygon**. Where the last point falls on the first the polygon is **closed**; otherwise it is *open*. Unless otherwise stated, the polygon is supposed to be closed. The points are the **vertices**, the tracts are the **sides** of the polygon. The closed polygon has the same number of vertices and sides, and we may call it an **n-angle** or **n-side**. The angles between the pairs of consecutive sides are the **angles** of the polygon, either inner or outer; unless otherwise stated, *inner* angles are referred to. Inner and outer angles at any vertex are supplemental. When each inner angle is less than a straight angle, the polygon is called **convex**; otherwise, *re-entrant*. Unless otherwise stated, convex polygons are meant. Sides and angles of a polygon may be reckoned either clockwise or counter-clockwise.

14. Prove that the sum of the inner angles of an n-side is $(n-2)$ straight angles. What is the sum of the outer angles?

15. Find the angle in a regular (*i.e.* equiangular and equilateral) 3-side, 4-side, 5-side, 8-side, 12-side. (For proof that there is a regular n-side, see Art. 137.)

16. Show that a (convex) polygon cannot have more than three obtuse outer angles, nor more than three acute inner angles.

17. Two angles of a $\triangle$ are α and β; find the angles at the intersection of their mid-rays.

18. If two $\triangle$ have their sides parallel or perpendicular in pairs, then the $\triangle$ are mutually equiangular.

19. The medial to the hypotenuse of a right $\triangle$ cuts the $\triangle$ into two isosceles $\triangle$.

20. An angle in a $\triangle$ is obtuse, right, or acute, according as the medial to the opposite side is less than, equal to, or greater than, half the opposite side.

EXERCISES I.

21. A medial will be greater than, equal to, or less than, half the side it bisects, according as the opposite angle is acute, right, or obtuse.

22. If P and Q be on the mid-normal of AB, then $\triangle APQ \equiv \triangle BPQ$ ($\equiv$ indicates congruence).

23. AB is the base, C the opposite vertex of an isosceles $\triangle$; show that $ABN \equiv BAM$ (1) when AM and BN are altitudes, (2) when they are medials, (3) when they are mid-rays of angles A and B, (4) when MN is normal to the mid-normal of AB.

24. P is any point within the $\triangle\ ABC$; show that $AP+BP<AC+CB,\ AP+PB+CP>\frac{1}{2}(AB+BC+CA)$.

25. $ABC\cdots L$ and $AB'C'\cdots L$ are two convex polygons, not crossing each other, between the same pairs of points, A and L; which is the longer? Give proof.

26. P is a point within $\triangle ABC$; show that angle $APB > ACB$ and sum of angles at $P = 2(A+B+C)$.

27. P is equidistant from A, B, and C; show that angle $APB = 2$(angle ACB).

28. Conversely, if angle $APB = 2$ (angle ACB), angle $BPC = 2$ (angle BAC), and angle $CPA = 2$ (angle CBA), then P is equidistant from A, B, C.

29. The mid-rays of the angles at the ends of the transverse axis of a kite cut the sides in the vertices of an anti-parallelogram (Art. 99).

30. The four joins of the consecutive mid-points of the sides of a 4-side form a parallelogram.

31. The joins of the mid-points of the pairs of opposite sides and of the pairs of diagonals of a 4-side concur, bisecting each other.

32. The mid-parallels to the sides of a △ cut it into 4 congruent △.

33. What figures are formed by the mid-parallels when the △ is right? isosceles? regular?

34. A parallelogram is a rhombus if a diagonal bisects one of its angles.

35. A parallelogram is a square if its diagonals are equal and one bisects an angle of the parallelogram.

36. From any point in the base of an isosceles △ parallels are drawn to the sides; the parallelogram so formed has a constant *perimeter* ($=$ measure round $=$ sum of sides).

37. The sum of the distances of any point on the base of an isosceles △ from the sides is constant.

38. The sum of the distances of any point within a regular △ from the sides is constant. — What if the point be without the △?

39. P is on a mid-ray of the angle A in the △ABC; compare the difference of PB and PC: when P is within the △, and when P is without.

40. The inner mid-ray of one angle of a △ and the outer mid-ray of another form an angle that is half the third angle of the △.

41. O is the orthocentre of the △ABC; express the angles AOB, BOC, COA, through the angles A, B, C.

42. Do the like for the circum-centre S and the in-centre I.

43. The medial to the hypotenuse of a right △ equals one-half of that hypotenuse.

44. The mid-rays of two adjacent angles of a parallelogram are normal to each other.

EXERCISES I. 75

45. In a 5-pointed star the sum of the angles at the points is a straight angle. What is the sum in a 7-pointed star?

46. Parallels are drawn to the sides of a regular △, trisecting the sides; what figures result?

47. A side of a △ is cut into 8 equal parts, through each section point parallels are drawn to the other sides; how are the other sides cut and what figures result?

48. Two △ are congruent when they have two mid-tracts of two corresponding angles equal, and *besides* have equal

(1) these angles and a pair of the including sides; or

(2) two pairs of corresponding angles; or

(3) one pair of corresponding angles and the corresponding angles of the mid-tract with the opposite side; or

(4) one pair of including sides and the adjacent segment of the opposite side.

49. Two △ are congruent when they have two corresponding sides and their medials equal, and *besides* have equal

(1) another pair of sides; or

(2) the angles of the medial with its side (in pairs); or

(3) a pair of angles of the bisected side with another side, the angles of the medial with this side being both acute or both obtuse; or

(4) a pair of angles of the medial with an including side, the corresponding angles of the medial with its side being both acute or both obtuse.

50. Two △ are congruent when they have a pair of corresponding altitudes equal, and *besides* have equal

(1) the pair of bases and a pair of adjacent angles; or

(2) the pair of bases and another pair of sides; or

(3) the pairs of angles of the altitude with the sides; or

(4) two pairs of corresponding angles; or

(5) the two pairs of sides, when the altitudes lie both between or both not between the sides of the △.

SYMMETRY.

90. We have seen that congruent figures are alike in size and shape, different only in place, and may be made to fit point for point, line for line, angle for angle. The parts that fit one on the other are said to **correspond** or be **correspondent**. Plainly only like can correspond to like, as point to point, etc.

Def. The ray through two points we may call the **join** of those points, and the point on two rays the **join** of the rays.

91. It is now plain that if A corresponds to A' and B to B', then the join of A and B must correspond to the join of A' and B'; for in fitting A on A' and B on B' the ray AB must fit on the ray $A'B'$ (why?). Also if the ray L corresponds to L', and M to M', then the join of L and M must correspond to the join of L' and M' (why?). These facts are very simple but very important.

We shall think of the plane as a thin double film, the one figure drawn in the upper layer, the other in the lower.

92. Two congruent figures may be placed anywhere and any way in the plane, but there are *two* positions especially important: (1) the one in which the one figure may be superimposed on the other by turning the one half of the plane through a straight angle about a ray called an **axis**; (2) the one in which the one figure may be fitted on the

SYMMETRY. 77

other by turning the one half of the plane through a straight angle about a point called a **centre**.

Congruent figures in either of these two positions are called **symmetric**: in the first case **axally,** as to the axis of symmetry; in the second case **centrally,** as to the centre of symmetry.

93. In two symmetrics, corresponding angles, like all other correspondents, are of course congruent; but they are reckoned *oppositely* if the symmetry be axal, *similarly* if

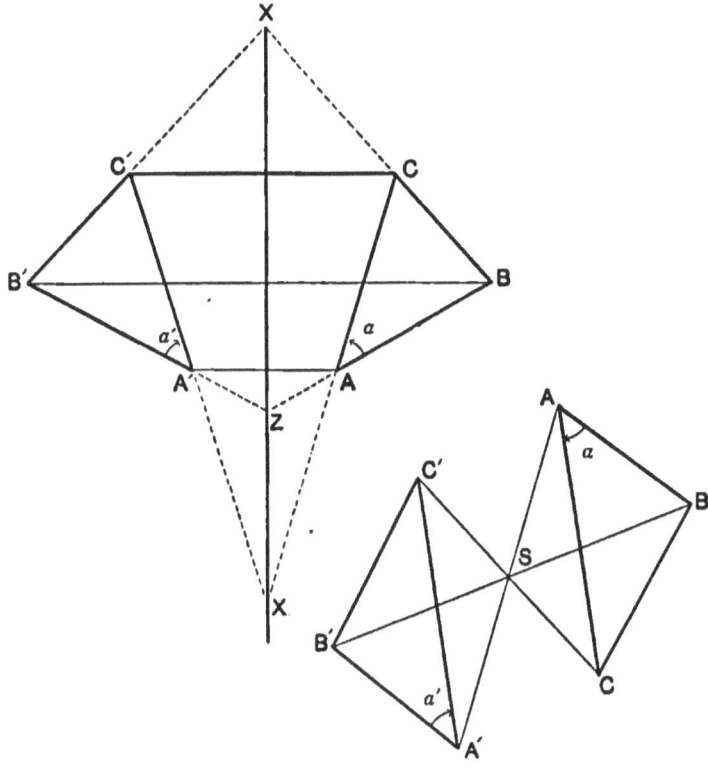

FIG. 61.

it be central. To parallels correspond parallels; to normals, normals; to mid-points, mid-points; to mid-rays, mid-rays; to the axis corresponds the axis, each point to itself; to the centre corresponds the centre itself (Fig. 61).

Elements, whether points or lines, that correspond to themselves may be called *self-correspondent* or **double**.

It is also manifest that centre and axis are the *only* self-correspondents; hence if a point be self-correspondent, it must lie on the axis in axal symmetry, *or* be the centre in central symmetry; and if two counter half-rays be correspondent, they (or the ray) must be normal to the axis in axal symmetry, *or* go through the centre in central symmetry.

94. These facts are all perfectly obvious, but a more vivid exemplification of the nature of these two kinds of symmetry may perhaps be found in the following:

Suppose the axis of symmetry to be a perfect plane mirror; then either half of the plane may be treated as the reflection or exact image of the other, and will be the symmetric of the other as to the mirror-axis. For the image of any point A is the point A' such that the axis is the mid-normal of AA', as we know from Physics; also, on folding over the one half of the plane about the axis upon the other half, the point A falls on A' (why?); hence A' is the symmetric of A as to the axis.

Suppose the centre of symmetry S to be also a reflector; then the reflection or image of any point A will be a point A' such that S is the mid-point of the tract AA', and on rotation through a straight angle about S the point A falls on A', and the half-ray SA fits on the half-ray SA'. Hence either of two centrally symmetric figures is the exact image of the other reflected from the centre of symmetry S.

Note carefully that these two species of symmetry depend upon the two fundamental definitive properties of the plane : central symmetry upon the *homœoidality* of the plane, axal symmetry upon the *reversibility* of the plane. Moreover, axally symmetric figures can *not* be fitted on each other without reversion, folding over ; by movement *in* the plane their corresponding parts can at best be *op*posed, but never *super*posed ; while on the other hand central symmetrics may be *super*posed, but cannot be *op*posed, along any ray, by motion in the plane. In central symmetrics the corresponding parts follow one another in the same order, but in axal symmetrics they follow in opposite orders.

95. We must now discuss these two symmetries more minutely, and to exhibit a certain remarkable relation holding between them we arrange their properties in parallel columns.

IN AXAL SYMMETRY.	IN CENTRAL SYMMETRY.
1. The axis corresponds to itself.	1. The centre corresponds to itself.
2. Every point of the axis corresponds to itself.	2. Every ray through the centre corresponds to itself (each half to the other).
3. Every self-correspondent point lies on the axis.	3. Every self-correspondent ray goes through the centre.
4. The join of two correspondent **rays** is on the axis.	4. The join of two correspondent **points** goes through the centre.
(For it is self-correspondent.)	(For it is self-correspondent.)
5. Correspondent **points** are equidistant from every **point** on the axis.	5. Correspondent **rays** are equally inclined (isoclinal) to every **ray** through the centre; hence **they are parallel**, as is otherwise manifest.

80   GEOMETRY.

6. The axis is a mid-**ray** of every **angle** between correspondent **rays**, and in fact the *inner* mid-ray. N.B. The *outer* mid-ray is a **normal to the axis**.	6. The centre is a mid-**point** of every **tract** between correspondent **points**, and in fact the *inner* mid-point. N.B. The *outer* mid-point is a **point at infinity**.
7. The join of two correspondent *points* is a **normal to the axis**.	7. The join of two correspondent *rays* is **at infinity**. (For they are parallel.)
8. Correspondent **tracts** are anti-parallel.	8. Correspondent **angles** are contra-posed (*i.e.* have their arms extended oppositely).
9. Correspondent **points** are equidistant from the axis.	9. Correspondent **rays** are equidistant from the centre.
10. The join of two **rays** and the join of their correspondents themselves correspond.	10. The join of two **points** and the join of their correspondents themselves correspond.

96. On regarding closely these correlated propositions, it becomes clear that the one set differs from the other only in the interchange of certain notions, as *point* and *ray*, *tract* and *angle*, etc. Every property of axal symmetry has its obverse in central symmetry, and *vice versa*. This most profound, important, and interesting fact has received the name of the **Principle of Reciprocity**. We make this notion more precise by the following

Def. *Two figures such that to every point of each corresponds a ray of the other, and to every ray of each a point of the other, are called* **reciprocal**. For example:

Suppose rays drawn through a point O to any number of points, A, B, C, D, E, . . . on a ray L. Then the point O with its ray through it, and the ray L with its point on it, are two reciprocal figures (Fig. 62). The first is called a (flat) **pencil** of rays, O being the centre; the second is called a **row** (or range) of points, L being the axis. Sup-

SYMMETRY. 81

pose we have now a second pencil through O' and a second row on L'. These two figures are again reciprocal, and the two pairs of reciprocals together make up another more complex pair of reciprocals. In this latter pair we find our definition fully exemplified. To O and O' correspond L and L'; to the rays through O and O' correspond the points on L and L'; also, to the join (ray) of O and O' corresponds the

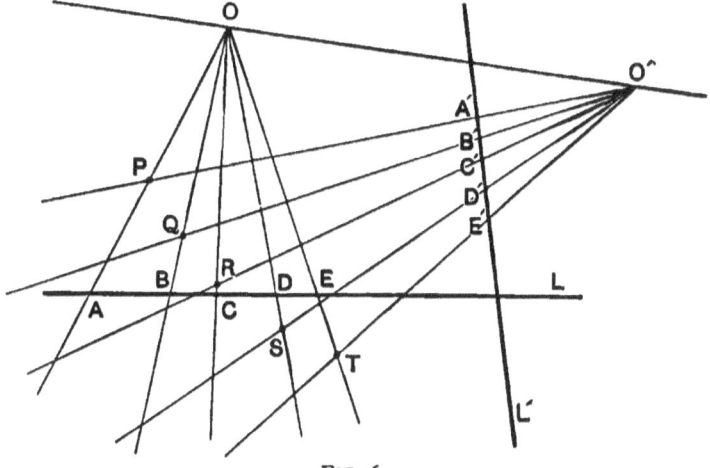

FIG. 62.

join (point) of L and L'; to any point as P, the join of two rays $(OA, O'A')$, corresponds a ray AA', the join of two points (A, A'). So Q, R, S, T are points corresponding to the rays BB', CC', DD', EE'. We may notice further that **angle** and **tract** correspond in the reciprocal figures; thus the angle AOB corresponds to the tract AB, and the angle BOC to the tract BC; while the angle OPO' corresponds to the tract AA' and the tract RS to the angle between the rays corresponding to R and S; namely, between CC' and DD'. Let the student trace out as many correspondences as possible.

97. To three points fixing a triangle in either of two reciprocals must correspond also three rays fixing a triangle in the other reciprocal; hence, in general, **triangle** corresponds to **triangle** in reciprocals. But notice: the *sides* of one correspond to the *vertices* of the other; hence if the sides of one all go through the same point, the vertices of the other all lie on the same ray; that is, **three concurrent rays** in either reciprocal correspond to **three collinear points** in the other.

It now appears that axal and central symmetry are reciprocal to each other; the reciprocal of an axal symmetric is a central symmetric, and the reciprocal of a central symmetric is an axal symmetric; the reciprocal properties of axal symmetry are the properties of central symmetry, and the reciprocal properties of central symmetry are the properties of axal symmetry.

Very often the two symmetric figures may be regarded as the two halves of one figure; this one figure is then said to be **symmetric** as to the axis of symmetry or as to the centre of symmetry, as the case may be.

98. If our figure be two points, A and A', then the midnormal X of the tract AA' is the axis of symmetry, manifestly. If, now, any double point D on the axis be joined with A and A', there results the isosceles $\triangle$ ADA', whence it appears that (Fig. 63)

The isosceles $\triangle$ is a symmetric $\triangle$.

It is plain that any two points on the ray AA' equidistant from N are symmetric as to X, that all points on the ray, and indeed in the whole plane, may be arranged in symmetric pairs, the members of each pair equidistant from the axis X.

99. Now take two points on the axis, as D and D', or D and D'', and consider the 4-side $DAD'A'$. It is composed of two △, ADD' and $A'DD'$, symmetric with each other as to the axis X, and opposed along that axis. Hence the 4-side is itself symmetrical as to X.

Def. Such a 4-side, with an axis of symmetry, is called a **kite**.

If we hold D fast, and let D' glide along X, the 4-side $ADA'D'$ remains a kite. We see that there are two kinds

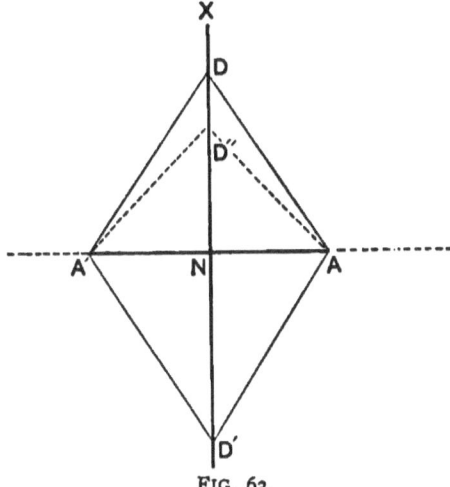

FIG. 63.

of kites, the convex kite, as $ADA'D'$, and the re-entrant, as $ADA'D''$. As the gliding point passes through N the kite changes from one kind to the other, passing through the intermediate form of the symmetrical △.

When the gliding point reaches a position D' such that $ND = ND'$, then the four sides of the kite are all equal (why?), and the kite becomes a **rhombus** (why?). In this case D and D' are symmetric as to AA' as an axis of symmetry. Hence *the rhombus has two axes of symmetry;* namely, its *two diagonals.*

In all cases the diagonals, AA' and DD', of the kite are normal to each other (why?).

100. Now consider a pair of points, B and B', symmetric as to the axis X (Fig. 64). Then X is mid-normal of BB'.

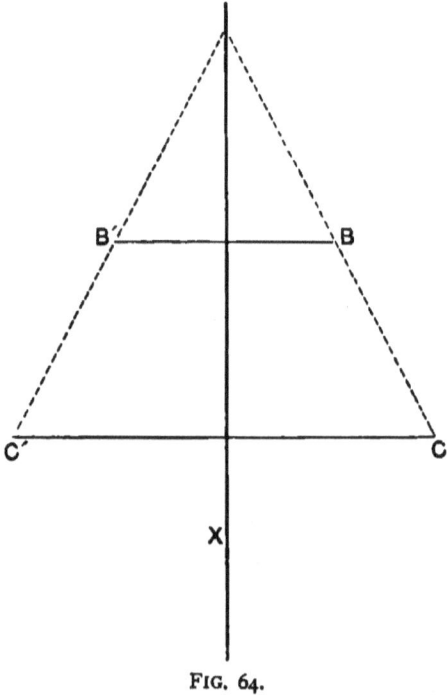

FIG. 64.

If C and C' be any other pair of symmetric points, then X is also mid-normal of CC'; hence BB' and CC' are parallel (why?). Also the tracts BC and $B'C'$ are symmetric as to X (why?), and the 4-side $BB'C'C$ is itself symmetric as to the axis X. Hence the angles at C and C' are equal,

SYMMETRY. 85

also the angles at B and B' are equal (why?); hence the angles at B and C and at B' and C' are supplemental (why?), and the 4-side $BB'C'C$ is an **anti-parallelogram** (why?). Hence we see that *another symmetric 4-side is an anti-parallelogram.*

It is plain that every anti-parallelogram is symmetric, for we know that the oblique sides prolonged yield an isosceles △. Let the student complete the proof.

101. There is *only one* kind of symmetric △, the isosceles. For, let ABA' (Fig. 65) be symmetric and A' correspondent

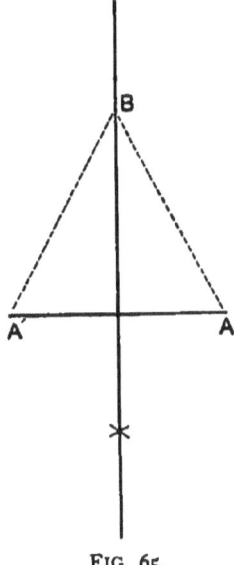

FIG. 65.

to A. Then B must correspond to itself (why?); hence B must lie on the axis (why?); hence $BA = BA'$ (why?). Now let the student prove that

(1) *In a symmetric △ the axis of symmetry is a medial;*
(2) *it is also a mid-ray;* (3) *it is also a mid-normal.*

GEOMETRY.

Conversely, let him show that

A medial that is a mid-ray, or a mid-normal, is an axis of symmetry.

102. There are *only two* axally symmetric 4-sides; namely, the kite and the anti-parallelogram. For, in a symmetric 4-side a vertex must correspond to a vertex (why?). Also, not all vertices can be on the axis (why?). Also, a vertex on the axis is a double point (why?). Also, the vertices not on the axis must appear in pairs (why?); hence there must be *either two or four* of them. If there be two only, then the other two are on the axis and the 4-side is a kite; if there be four of them, we have just seen that the 4-side is an anti-parallelogram.

103. Now let us turn to the *reciprocals.* The reciprocals of the two *points A* and *A'* symmetric as to the *axis X* will be two *rays L, L',* symmetric as to the *centre S.* But rays symmetric as to a centre are parallel (why?); hence we have two parallels symmetric as to *S*, which is midway between them. The rays are symmetric as to any other point *S'* midway between them (why?). The piece of plane between these parallels is called a *parallel strip,* or **band** (Fig. 66).

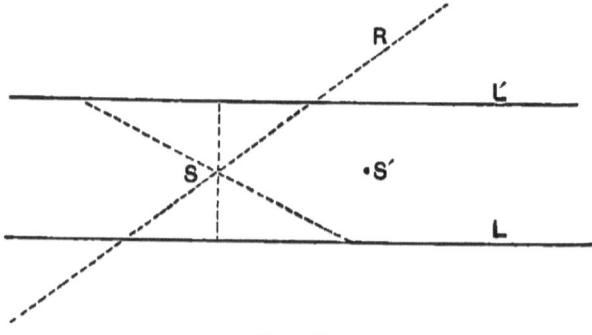

FIG. 66.

But what corresponds to the point D on the axis X? The answer is: a ray R through S (why?). Hence to the symmetric $\triangle$ of the three points A, A', D, there corresponds the figure formed by two parallels L, L', and a transverse R through S,—a so-called *half-strip*. This is truly a *three-side*, but not apparently a $\triangle$ (3-angle), for the parallels do not meet in finity, in regions accessible to our experience. Hence, instead of saying that the reciprocal of a $\triangle$ in axal symmetry is a $\triangle$ (3-angle or 3-point) in central symmetry, we should have said, accurately, that the reciprocal of a $\triangle$ in axal symmetry is a 3-side (or trilateral) in central symmetry, which will always be a $\triangle$ except when sides are parallel or all concur. In higher Geometry it is very convenient to remove this apparent exception by using this form of expression: the parallels meet *not in finity*, but *in infinity*.

104. It is indeed plain that

A $\triangle$ can have no centre of symmetry.

For, since vertex corresponds to vertex, and since correspondents appear in pairs, one vertex must be a double point; hence it would have to be the centre S (why?). But the other two vertices would have to lie on a ray through S, being correspondents; hence the three vertices would be collinear, and the $\triangle$ would be flattened out to a *triply-laid ray*.

105. But there is a centrally symmetrical 4-side; namely, the **parallelogram**. For, consider once more the kite $AXA'X'$ and let us reciprocate it into a centrally symmetric figure (Fig. 67). To the axis XX' will correspond the centre S; to the symmetric pair of rays AX and $A'X$ will correspond a symmetric pair of points P and P'; to the join of those on the axis X will correspond the join of these through

88 GEOMETRY.

the centre (PP'). Similarly, to the symmetric rays AX' and $A'X'$ will correspond the symmetric points Q and Q', and to the join X' will correspond the join QQ'. Also, AX and AX' have a join A while $A'X$ and $A'X'$ have a join A', and these joins are symmetric as to the axis XX'; recipro-

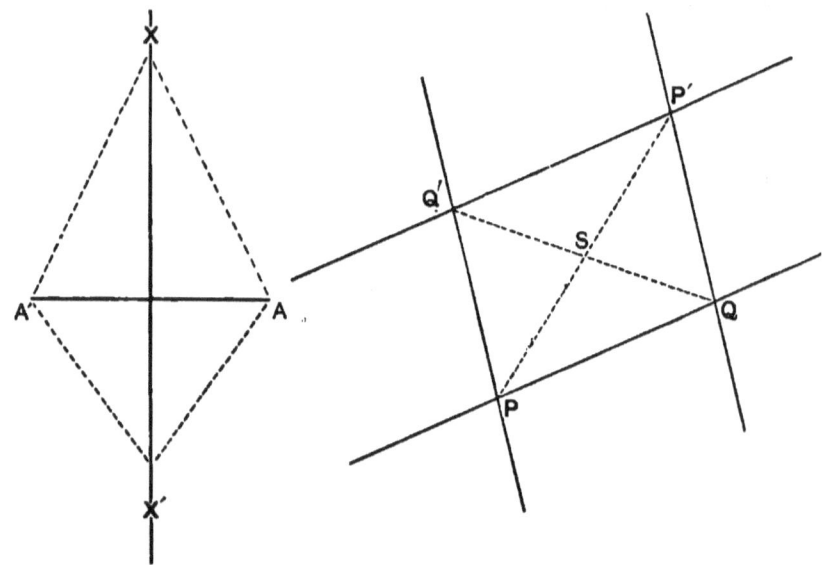

FIG. 67.

cally, P and Q have a join PQ, and P' and Q' have a join $P'Q'$, and these joins are symmetric as to S; that is, they are parallel (why?). Similarly, PQ' and $P'Q$ correspond to B (AX, $A'X'$) and B' ($A'X$, AX'); but B and B' are symmetric as to XX' (why?); hence PQ' and $P'Q$ are symmetric as to S, *i.e.* are parallel. Hence $PQ P'Q'$ is symmetric as to S, and is a *parallelogram*. Q. E. D.

106. We may indeed see at once that since any two parallels are centrally symmetrical as to any mid-point, a pair

of parallels or a parallelogram is symmetric as to the common mid-way point, the intersection of the diagonals. But the foregoing reciprocation is instructive, as illustrating in detail the method to be pursued, and as showing the intimate relation of the different symmetric quadrilaterals; namely, *the* **parallelogram** *is the common reciprocal of both* **kite** *and* **antiparallelogram**, which are thus seen to be really *one*.

107. Central symmetry does not in general imply anything at all with respect to axal symmetry in a figure. We may draw through any point S any number of rays and lay off on each from S a pair of counter tracts SP and SP', SQ and SQ', etc. No matter how PQ, etc., be chosen, the figure so obtained will be centrally symmetric as to S; but it may have no axal symmetry whatever. Neither does axal symmetry in general imply any central symmetry, but we may establish the following important

Theorem. — *Any figure with two rectangular axes of symmetry has also a centre of symmetry; namely, the intersection of those axes.*

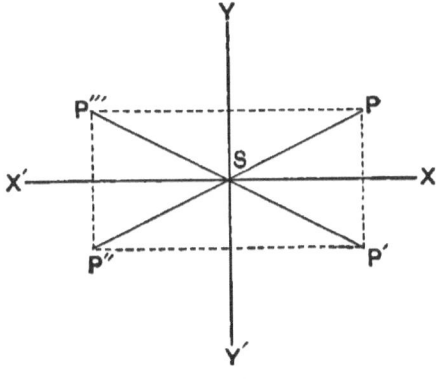

FIG. 68.

Data: XX' and YY' two rectangular axes, P any point of a figure symmetric as to these axes (Fig. 68).

Proof. The point P' symmetric with P as to XX' is a point of the figure (why?); also P'' symmetric with P' as to YY' is a point of the figure (why?); so too is P''' (why?); the figure $PP'P''P'''$ is a rectangle (why?), its diagonals halve each other, and $SP = SP'' = SP' = SP'''$. Hence S is a centre of symmetry.   Q. E. D.

THE CIRCLE.

108. We have already discovered the existence of a *homœoidal* plane curve *not reversible* and have named it **circle**.

Defs. A ray cutting a curve is called a **secant**, as L; the part of the secant intercepted by the curve, or the tract between two points of the curve, is called a **chord**, as AB. A finite part of a curve is called an **arc**. A chord and an arc with the same two ends are said to **subtend** each other. Also, the intercept of any line between the ends of an angle is said to *subtend* the angle. Thus BC and DE subtend the angle O (Fig. 69).

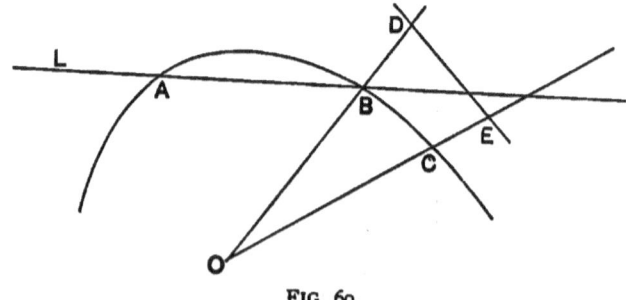

FIG. 69.

109. Theorem XLVIII. — *Congruent arcs subtend congruent chords.*

Proof. Let the arcs AB and CD be congruent; then we may fit A on C and at the same time B on D; then the chords AB and CB fit throughout (why?).   Q. E. D.

N.B. We can *not* convert this proposition at once (why?) (Fig. 70).

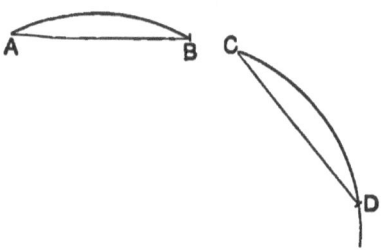

FIG. 70.

110. Theorem XLIX. — *A closed curve is cut by a ray in an even number of points* (Fig. 71).

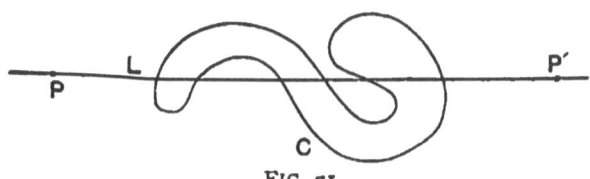

FIG. 71.

Proof. Let L be a ray, C any closed curve. Suppose a point P to trace out the ray L. At first P is without the curve, at last it is also without the curve; hence P has crossed the curve going out as often as it has crossed the curve going in, for every entrance there is an exit; hence the points of intersection appear in pairs, their number is even, as 0, 2, 4, 6, . . . $2n$.   Q. E. D.

These preliminary or auxiliary theorems, which prepare the way for a theorem to follow, are sometimes called **lemmas** (λημμα = assumption, premise, support, prop).

*111. **Theorem L.** — *A circle has an axis of symmetry through every one of its points* (Fig. 72).

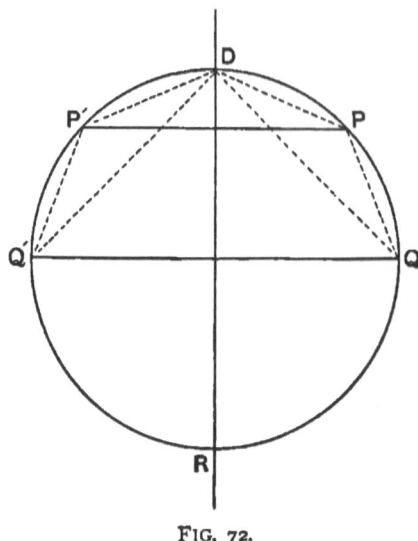

FIG. 72.

Proof. Let D be any point of a circle. Take any arc DP, and slip it round till P falls on D and D on P'; this is possible (why?). Then PDP' is a symmetrical $\triangle$ (why?); and its axis of symmetry DR halves normally the chord PP', and also halves the angle PDP' (why?). Now take any other arc DQ and slip it round till Q falls on D and D on Q', so that DQ and $Q'D$ are congruent. Then the chords DQ and DQ' are congruent (why?). Also, on taking away the congruents DP and DP' we have left PQ and $P'Q'$ as congruent remainders. Hence the chords PQ

and PQ' are congruent (why?). Hence the △ PDQ and $P'DQ'$ are congruent (why?); hence the angles PDQ and $P'DQ'$ are equal (why?); hence DR halves also the angle QDQ' (why?). But the △ QDQ' is symmetric (why?); hence DR is also its axis of symmetry, and Q and Q' are symmetric points of the circle; hence any point of the circle has its symmetric point as to DR; *i.e.* DR is an axis of symmetry of the circle. Moreover, D was any point of the circle; hence through any point of the circle passes an axis of symmetry. Q. E. D.

Def. A ray halving a system of parallel chords is called a **diameter**; the chords and diameter are called **conjugate** to each other.

Corollary 1. In a circle a diameter is normal to its conjugate chords.

Corollary 2. Every mid-normal to a chord in a circle is a diameter and halves the subtended arcs.

*112. **Theorem LI.** — *A circle has a centre of symmetry* (Fig. 72).

For the ray through D must cut the circle in some second point, as R (why?), and as the ray turns round from the position DR to the reversed position RD, through a straight angle, it must pass through some position, QQ', normal to its original position (why?). Hence for any axis of symmetry there is another normal thereto and their intersection is a centre of symmetry (why?). Q. E. D.

N.B. There is *only one* centre of symmetry (why?).

Def. This centre of symmetry is named **centre** *of the circle*. It is often convenient to call the whole ray through the centre a *centre ray or line*, and to restrict the term *diameter* to the centre chord.

Corollary 1. All diameters go through the centre, and halve each other there; *conversely*, chords halving each other are diameters.

Def. Two diameters each halving all the chords parallel to the other are called **conjugate**.

Corollary 2. In the circle two diameters normal to each other are conjugate; and *conversely*, two conjugate diameters are normal to each other.

N.B. Other curves, as Ellipse and Hyperbola, have conjugate diameters *not in general* normal to each other (Fig. 73).

*113. **Theorem LII.** — *All diameters of a circle are equal* (Fig. 74).

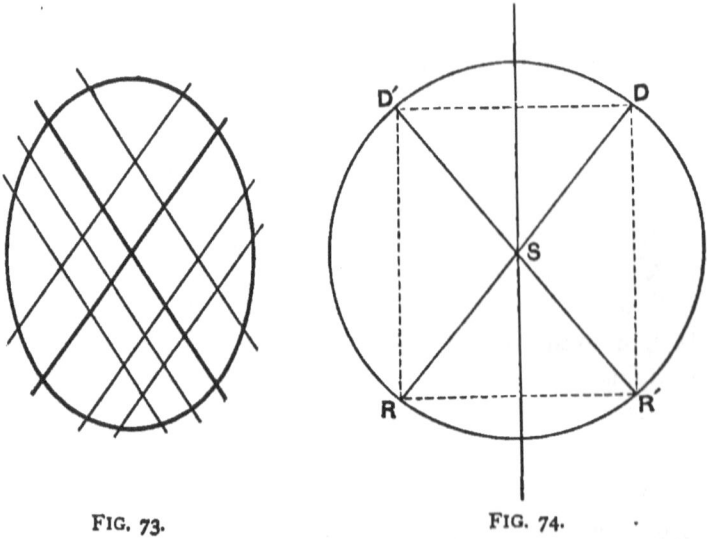

FIG. 73.   FIG. 74.

Proof. Let DR and $D'R'$ be two diameters. The figure $DD'RR'$ is a parallelogram (why?), and DD' is parallel to RR'; hence the mid-normal of these parallels is a diameter through the centre S; hence SD and SD' are symmetric and equal; hence $DR = D'R'$. Q. E. D.

Def. A half-diameter, from centre to circle, is called a **radius**.

Corollary 1. All radii of a circle are equal; or, all points of a circle are equidistant from the centre.

Corollary 2. Every parallelogram inscribed in a circle is a rectangle.

N.B. By help of this important property the circle is commonly defined as *a plane curve all points of which are equidistant from a point within called the centre.* The common distance of all points of the circle from the centre is often called **the radius**. We have deduced this property from the homœoidality; *conversely,* we may deduce the homœoidality from this property taken as definition. But if there were no such surface as the plane, at least for our intuition, the circle might still exist on the sphere-surface, without centre, but with the body of its properties unimpaired. Hence it seems better to define the circle by its intrinsic homœoidality than by its extrinsic centrality.

Corollary 1. All points within the circle are less, and all points without are more, than the radius distant from the centre.

Defs. The two symmetric halves into which a diameter cuts a circle are called **semicircles**. The part of the plane bounded by an arc and its chord is called a **segment**; the part bounded by an arc and the two radii to its ends is called a **sector**. If the sum of two arcs be a circle, we may call them **explemental**, the one *minor*, the other *major;* every chord belongs equally to each of two explemental arcs, but in general, unless otherwise stated, it is the *minor* that is referred to. Two arcs whose sum is a half-circle are called **supplemental**; two whose sum is a quarter-circle or quadrant are called **complemental**.

Corollary 2. All circles of the same radius are congruent; also, all semicircles of the same radius are congruent, and all quadrants of the same radius are congruent.

Corollary 3. Any circle may be slipped round at will upon itself about its centre as a pivot, like a wheel about its axle, without changing in the least the position of the whole circle.

114. From the foregoing it is clear that if we hold one point of a ray fixed, and turn the ray in the plane about the fixed point, every other point of it will trace out a circle about the fixed point as a centre. An instrument, one point of which may be fixed while the other is movable about in a plane, is called a **compass** or pair of **compasses**, and is both the simplest and the most important of all instruments for drawing.

115. **Theorem LIII.** — *Through any three points not collinear one, and only one, circle may be drawn.*

Proof. Let A, B, C be the three points not collinear (Fig. 75). We have already seen that the mid-normals to the tracts AB, BC, CA concur in a point S equidistant from A, B, and C; hence a circle about S with radius d passes through A, B, C. Also there is only one point thus equidistant from A, B, C (why?); hence there is only one circle through A, B, C. Q. E. D.

Def. The circle through the vertices A, B, C, of a $\triangle$ is called the **circum-circle** of the $\triangle$.

Corollary 1. A $\triangle$, or a triplet of points, or a triplet of rays, determines one, and only one, circle.

THE CIRCLE.

Corollary 2. Through two points, A and B, any number of circles may be drawn. Their centres all lie on the mid-normal of AB.

Corollary 3. As BC turns clockwise about B as a pivot, the intersection S, the centre of the circle through A, B, C, retires upward ever faster and faster along the mid-normal N of AB; when C becomes collinear with A and B, the inter-

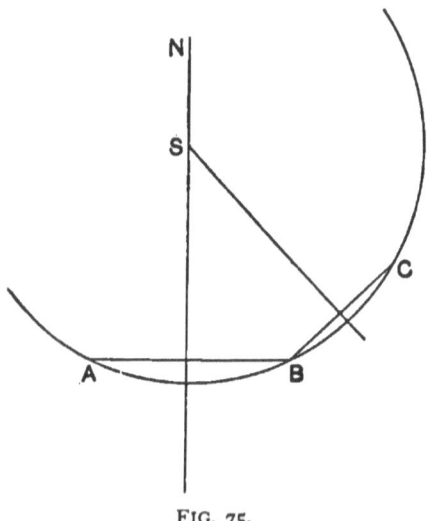

FIG. 75.

section of the mid-normals of AB and BC vanishes from finity, or *retires to infinity*, as the phrase is. As BC keeps on turning, S reappears in finity below and moves slower and slower upward along the mid-normal. Moreover, a circle passes through A, B, and C, no matter how close C may lie to the ray AB, nor on which side of it : only as C falls upon the ray does the centre of the circle vanish into infinity; that is, we may draw a circle that shall fit *as close to the ray AB as we please*, though not upon it, by retiring the centre far enough. Hence a ray may be conceived as a circle with centre retired to infinity; it is strictly the **limit** of a circle whose centre has retired, along a normal to it, *without limit*.

116. Theorem LIV. — *A circle can cut a ray in only two points.*

For there are only two points on a ray at a given distance from a fixed point (why?). Q. E. D.

117. Theorem LV. — *Secants that make equal angles with the centre ray* (or axis) *through their intersection intercept equal arcs on the circle.*

Proof. For both the two semicircles and the two secants are symmetric as to the axis *IS* (why?); hence, on folding over the one half-plane upon the other, A falls on A', B on B', arc a fits on arc a', and chord c on chord c' (Fig. 76). Q. E. D.

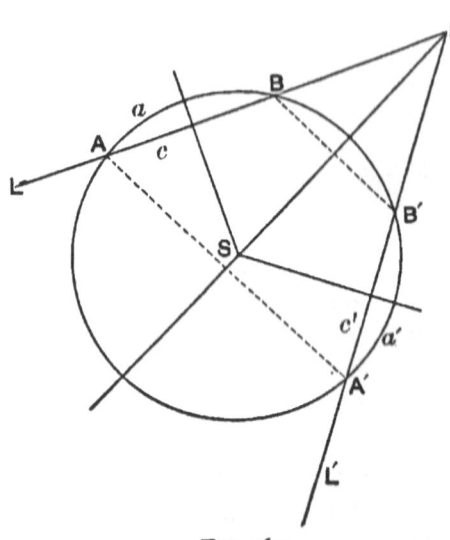

FIG. 76.

Conversely, *Secants that intercept equal arcs make equal angles with the axis through their intersection.*

Proof. Let L and L' intersect equal arcs AB and $A'B'$. Draw the mid-normal of AA'; it is an axis of symmetry (why?). On folding over the left half-plane upon the right half-plane, A falls on A' and B on B' (why?); hence AB and $A'B'$ are symmetric; hence they meet on the axis and make equal angles with it (why?). Q. E. D.

Corollary 1. Equal chords are equidistant from the centre; and *conversely*, Chords equidistant from the centre are equal.

Corollary 2. The greater of two unequal chords is less distant from the centre.

Corollary 3. A diameter is the greatest chord.

Corollary 4. Arcs intercepted by two parallel chords are equal.

Corollary 5. Equal chords or arcs subtend equal *central angles* (angles at the centre), and *conversely*.

Corollary 6. Of two unequal chords or arcs, the greater subtends the greater central angle.

What figure is determined by two parallel chords and the chords of the intercepted arcs? By two secants that intercept equal arcs and the central normals thereto?

118. Theorem LVI. — *A central angle subtended by a certain arc* (or chord) *is double the peripheral angle subtended by the same* (or an equal) *arc* (or chord) (Fig. 77).

Proof. Let ASB be a central angle, and APB be a *peripheral* angle (periphery = circumference, the circle itself), subtended by the same arc or chord AB. Draw the diameter PD. Then the △ ASP and BSP are isosceles (why?); hence the angle $ASD = 2$ angle APD, and angle $BSD = 2$ angle BPD (why?); hence angle $ASB = 2$ angle APB. Q. E. D.

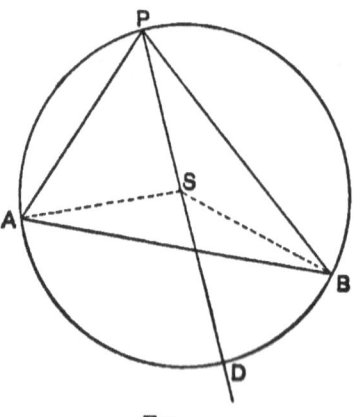

FIG. 77.

Corollary 1. All peripheral angles subtended by (or standing on) the same or equal chords or arcs are equal. Hence, as P moves round from A to B, the angle APB remains unchanged in size.

Def. An angle with its vertex on a certain arc, and its arms passing through the ends of that arc, is said to be **inscribed** in that arc. Hence for an angle to be *inscribed* in a certain arc, and for it to *stand on* the *explemental* arc, are equivalent.

Corollary 2. All angles inscribed in the same or equal arcs of the same or equal circles are equal.

Corollary 3. As the vertex P of a peripheral angle subtended by an arc (or chord) AB, in passing round a circle goes through either end of the arc (or chord), the angle itself leaps in value, changes to its *supplement*.

119. Theorem LVII. — *The locus of the vertex of a given angle standing on a given tract is two symmetric circular arcs through the ends of the tract* (Fig. 78).

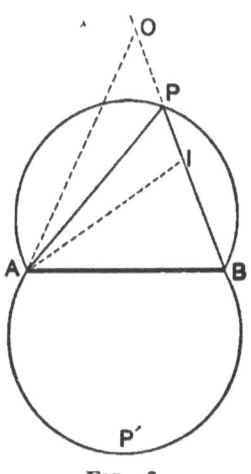

FIG. 78.

Proof. Let P be the vertex of the given angle, in any position, standing on the tract AB. Through A, P, and B draw a circular arc subtended by AB. We have just seen that as long as P stays on this arc, the angle P remains the same in size. Moreover, the point P cannot be without the arc, as at O, because the angle AOB is less than APB (why?); neither can it come within the arc, as to I, because the angle AIB is greater than APB (why?); hence so long as the angle is constant in size the vertex must remain on the arc APB or on its symmetric arc $AP'B$, of which plainly the same may be said.   Q. E. D.

120. Theorem LVIII. — *The angle inscribed in a semicircle* (or standing on a semicircle or diameter) *is a right-angle* (Fig. 79).

Proof. Let ABC be any angle in a semicircle. Then it is half of the central angle ASC (why?), which is a straight angle (why?). Q. E. D.

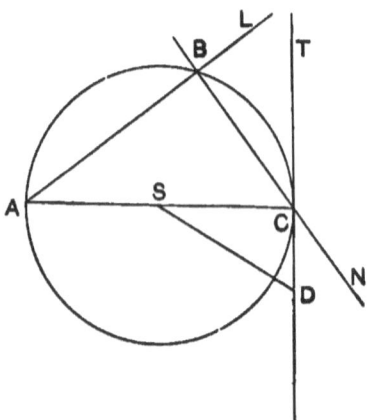

FIG. 79.

Now let the vertex B, the intersection of the rays L and N, move round the circle toward C; the angle ABC remains a right angle, *no matter how close B approaches to C;* moreover, when B passes C, into the lower semicircle, the angle remains a right angle (why?). That is, the angle at B remains a right angle, *no matter from which side nor how close B* approaches to C. Hence it is a right angle *even when B falls on C*. But then the ray L falls on the diameter AC, hence the ray N takes the position T normal to the diameter (or radius) at its end. Such a normal to a radius at its end is called a **tangent** to the circle at the point of tangence (or *touch* or *contact*) C.

Def. A ray normal to a tangent to a curve at the point of touch is called **normal** to the curve itself. Hence

Corollary. All radii of a circle are normal to the circle; and *conversely*, all normals to a circle are radii of the circle.

121. Theorem LIX. — *All points on a tangent, except the point of contact, lie outside of the circle.*

Proof. For the point of touch is distant radius from the centre (why?), and all other points, as D, of the tangent are further from the centre (why?); hence all other points of the tangent are without the circle (why?). Q. E. D.

122. Theorem LX. — *The point of tangence is a double point.*

Proof. For it is on a diameter, or axis of symmetry, of the circle, and every such point is a double point with respect to that axis.

Independently of this consideration, it is seen that the chord CB becomes the tangent CT *when, and only when,* the points B and C fall together in C.

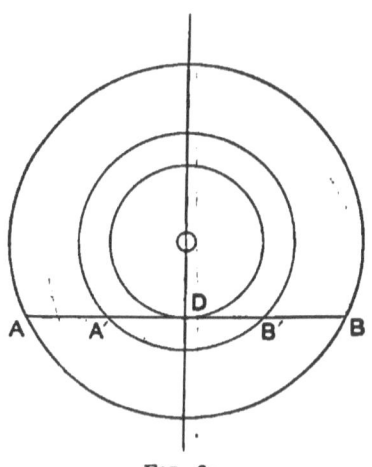

FIG. 80.

Still otherwise, let AB be any chord of a circle about (Fig. 80) O. Draw the mid-normal OD. Now let the circle shrink about the centre O: the points A and B move towards each other, and as D is always mid-way between them they finally fall together in D, and their join is tangent at D to the circle of radius OD.

Def. Two points thus falling together in a double point are called **consecutive** points. Accordingly we may define a tangent to a circle (or to any curve) as *a ray through two consecutive points of the circle* (or curve). Adopting this definition, let the student prove

123. Theorem LXI. *Every tangent to a circle is normal to a radius at its end;* conversely, *Every normal to a radius at its end is tangent to the circle.*

124. Theorem LXII. *The angle between a tangent and a chord equals the peripheral angle on the same chord, or equals half the angle of the chord* (Fig. 81).

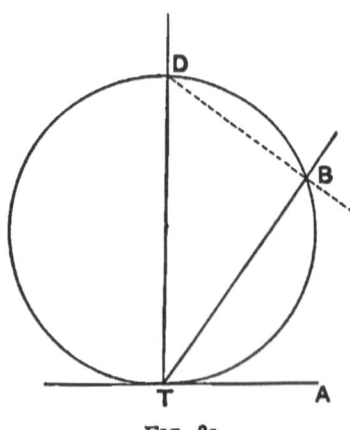

FIG. 81.

Proof. For if DT be a diameter, then the angles BDT and BTA are equal, being complements of the same angle BTD (why?). Or thus: TB is a chord, and TA is also a chord, through the double point T; hence the angle BTA is a peripheral angle standing on the arc TB. Q. E. D.

125. Theorem LXIII. — *The angle between two secants is half the sum or half the difference of the angles of the intercepted arcs, according as the secants intersect within or without the circle.*

Proof. For on drawing AB' the angle I is seen (Fig. 82) to be the sum, and the angle O the difference, of the

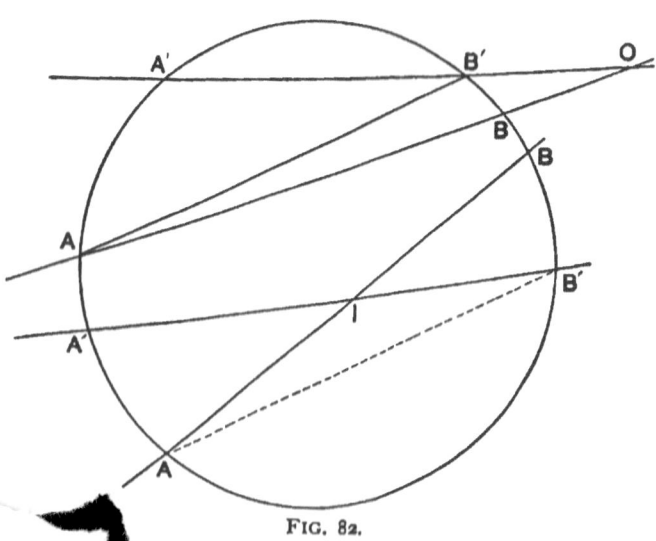

FIG. 82.

angles at A a[n]d B' standing on the arcs AA' and BB'.

Q. E. D.

126. Theorem LXIV. — *An encyclic quadrangle has its [opposite angles sup]plemental.*

[Proof. The] angles B and D are halves of the two [arcs CSA a]nd CSA, whose sum is a round angle. [Hence B a]nd D is a straight angle. Q. E. D.

127. Theorem LXV. — Conversely, *A quadrangle with its opposite angles supplemental is encyclic* (Fig. 83).

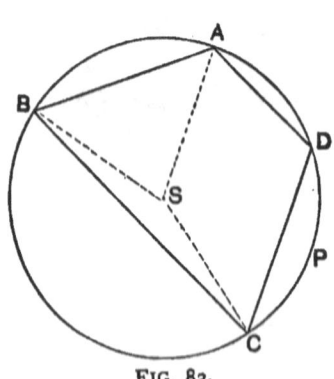

FIG. 83.

Proof. Let $ABCD$ be the quadrangle with the angles A and C, B and D, supplemental. About the $\triangle ABC$ draw a circle. If P be any point on the arc of this circle explemental to ABC, then the angle APC is the supplement of ABC; but if P be not on this arc, then the angle APC is either greater or less than that supplement (why?). Now the angle D is that supplement; hence D is on the arc. Q. E. D.

128. Relations of circles to each other.

Suppose two circles K and K' of radii r and r' concentric, *i.e.* to have the same centre O. Then, pla[...] the distance between them measured on a[ny half-axis AT,] is $r - r'$, the difference of the radii. Dr[aw tangents $r - r'$;] $A'T'$, where OO' cuts the circles. They a[re parallel (why?).] Now let the centre of K' move out on OO[' a distance $r - r'$;] then A falls on A' and $A'T'$ on AT; [the circles have a] *common tangent* at A and are said [to touch each other] *innerly* at A (Fig. 84).

Now let O' move still further along OO'; then the circles will lie partly within, partly without, each other; they will intersect at two points, and only two (why?), symmetric as to OO' (why?), namely P and P'; hence

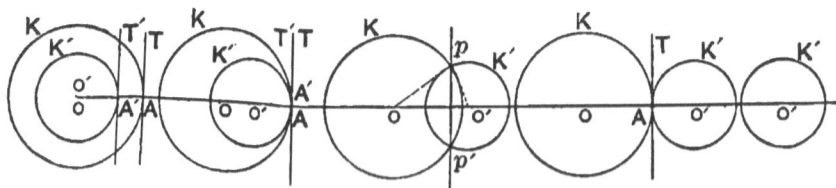

FIG. 84.

Theorem LXVI. — *The common axis of two circles is the mid-normal of their common chord.*

When O' is distant $r + r'$ from O, the circles lie without each other, but still have a common tangent (why?) and are said to *touch* **outerly**.

As O' moves still further away from O, the circles cease to touch and henceforth lie entirely without each other.

Thus we find that there are *three critical* positions depending on the distance d between the centres O and O':

$d = o$, when the circles are *concentric*.

$d = r - r'$, when the circles *touch innerly*.

$d = r + r'$, when the circles *touch outerly*.

There are also *three intermediate* positions:

For $o < d < r - r'$ the one circle is *within* the other.

For $r - r' < d < r + r$ the circles *intersect*.

For $r + r < d < \infty$ the circles lie *without* each other.

129. Theorem LXVII. — *From any point without a cir two, and only two, tangents may be drawn to the circle* (Fi 85).

108 GEOMETRY. [Th. LXVII.

Proof. Let O be the centre of the circle K, and P be the point without. On OP as a diameter draw a circle K'; only one such circle is possible (why?), and it cuts K in two, and only two, points, T and T'. Draw PT and PT': they are tangent to K at T and T' (why?). Moreover, no other ray through P, as PU, is tangent to K, because OUP is not a right angle (why?). Q. E. D.

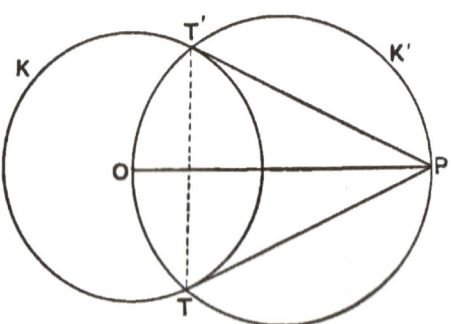

FIG. 85.

Def. The chord TT' through the points of contact of the tangents is called the chord of contact for the point P or the **polar** of the **pole** P (see Art.   ).

The angle between the tangents to two curves at the intersection of the curves is called the **angle between the curves** themselves. When this is a right angle, the curves are said to intersect **orthogonally**.

The distance PT or PT' is called the **tangent-length** from P to the circle.

Corollary 1. Two circles, one having as radius the tangent-length from its centre to the other, intersect orthogonally.

Corollary 2. Two tangents are symmetric as to the axis through their intersection; hence, also, the tangent-lengths are equal.

TH. LXIX.] THE CIRCLE. 109

130. Theorem LXVIII. — *All tangent-lengths to a circle from points on a concentric circle are equal, and intercept equal arcs of the circle* (Fig. 86).

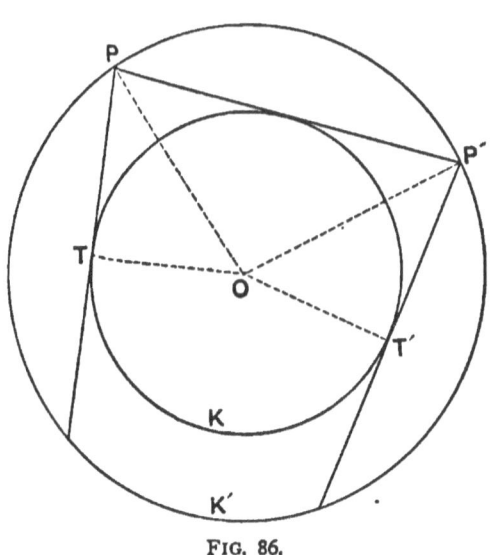

FIG. 86.

Proof. For if P be any point without the circle K', we may turn P round about the centre O on a concentric circle K' without affecting any of the relations obtaining (why?).

Or thus: the right $\triangle$ TOP and $T'OP'$ are plainly congruent (why?); hence $PT = P'T'$ (why?). Q. E. D.

*131. **Theorem LXIX.** — *The intercept between two fixed tangents on a third tangent subtends a constant central angle* (Fig. 87).

Proof. Let PT and PT' be the fixed tangents, VV' the intercept on the variable ray tangent at U. Then TPT' is a constant angle, and VOV' is half of TOT' (why?), and hence is constant. Q. E. D.

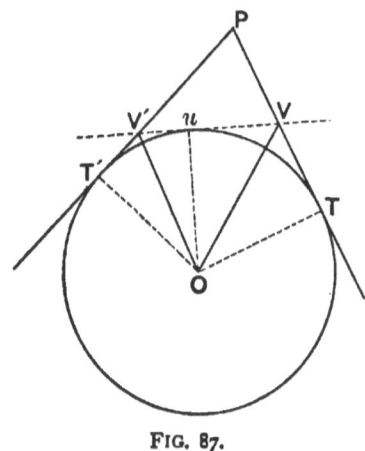

FIG. 87.

132. Theorem LXX.—*If the central* (or peripheral) *angles of the common chord of two intersecting circles be equal, the circles are equal.*

Let the student conduct the proof suggested by the figure (Fig. 88), and let him prove the *converse*.

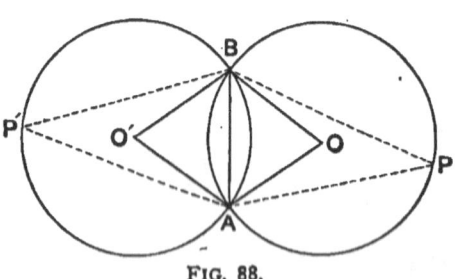

FIG. 88.

*133. **Theorem LXXI.** — *The circumcircle of a △ equals the circumcircle of the orthocentre and any two vertices of the △* (Fig. 89).

Proof. Let K be the circumcircle of the △ ABC, K' the circumcircle of A, B, and O the orthocentre. The angles

C and $B'OA'$ are supplemental (why?); also the angles D and BOA are supplemental (why?); and the angles BOA and $B'OA'$ are equal (why?); hence the angles D and C are equal; hence $K = K''$ (why?)   Q. E. D.

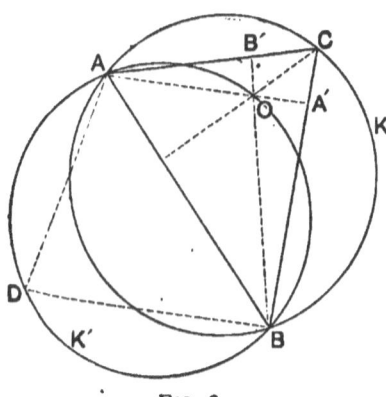

FIG. 89.

*134. **Theorem LXXII.**—*The mid-points of the sides of a $\triangle$, the feet of its altitudes, and the mid-points between its orthocentre and vertices, are nine encyclic points.*

Proof. Let a circle through X, Y, Z, the mid-points of the sides, cut the sides in three other points, U, V, W. Then the angle $ZXY =$ angle A (why?), and also $=$ angle ZVY (why?); therefore the $\triangle$ AZV is symmetrical. Hence the $\triangle$ ZVB is also symmetrical, Z is equidistant from A, V, and B, and the angle AVB is a right angle (why?); so also the angles at U and W; *i.e. the circle through the mid-points of the sides goes through the feet of the altitudes* (Fig. 90).

Again, if the circle cuts the altitudes at P, Q, R, then the angle $VPW =$ angle VZW (why?) $= 2$ angle VAW (why?). Moreover, A, V, O, W, are encyclic (why?); hence AO is a diameter of the circle through them (why?); and VAW is a peripheral angle standing on the arc VW; hence the double angle *VPW* must be the central angle of the same arc; *i.e. P* is the mid-point between a vertex and orthocentre : so, also, are *Q* and *R*, similarly.   Q. E. D.

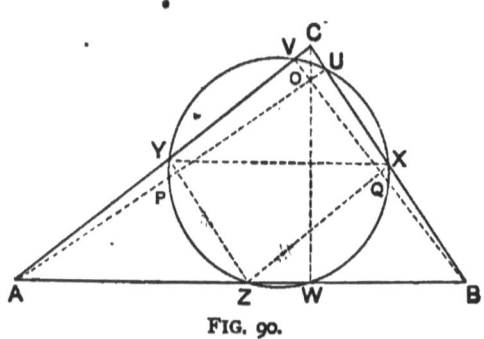

FIG. 90.

Def. This remarkable circle is called the **9-point circle**, or *circle of* **Feuerbach**, of the △ *ABC*.

Corollary. The radius of the 9-point circle is half the radius of the circumcircle.

135. *Def.* A **Polygon** all of whose sides touch a circle is said to be **circumscribed** about it, and the circle is said to be **inscribed** in the polygon.

Theorem LXXIII. — *A circle may be inscribed in any* △.

Proof. Let *ABC* be any △ (see Fig. 59). Draw the inner mid-rays of the angles at *A, B, C*; they concur in the in-centre *I* of the △, equidistant from the three sides (why?). About this point as centre with this common distance as radius draw a circle; it will touch the three sides of the △ (why and where?).   Q. E. D.

N.B. We have seen that the outer mid-rays of the angles concur in pairs with the inner mid-rays of the angles in the three **ex-centres** E_1, E_2, E_3; also equidistant from the sides

(Fig. 60). The circles about these touch only two sides innerly, but the third side outerly, and hence are called **escribed**, or **ex-circles**.

Corollary. Four, and only four, circles touch, each, all the sides of a △.

135 a. Theorem LXXIV. — *In a 4-side circumscribed about a circle the sums of the two pairs of opposite sides are equal* (Fig. 91).

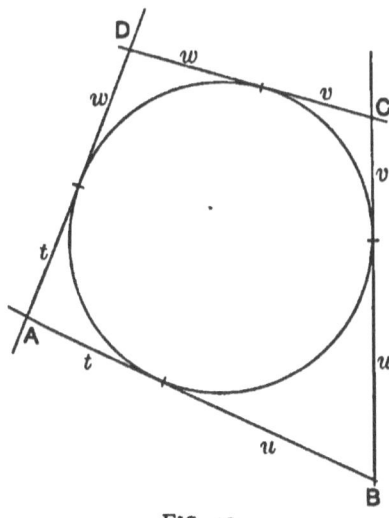

FIG. 91.

Proof. The sum of the four sides is plainly $2t + 2u + 2v + 2w$, and the sum of either pair of opposites is $t+u+v+w$.
Q. E. D.

Conversely, *If the sums of two pairs of opposite sides of a 4-side be equal, the 4-side is circumscribed about a circle.*

Proof. Let two counter sides, AB and DC meet in I, and inscribe a circle K in the triangle ADI. Through B

114   GEOMETRY.   [TH. LXXV.

draw a tangent (Fig. 92) to K at U, and let it cut DI at C'. Then since $ABC'D$ is circumscribed about K, we have

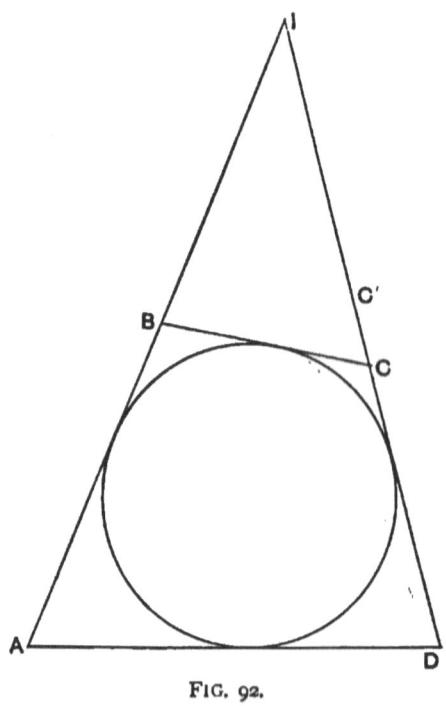

FIG. 92.

$$AB + C'D = BC' + DA.$$
Also $\quad AB + CD = BC + DA$ (why?).
Whence $\quad CD - C'D = BC - BC',$
or $\quad CC' = BC - BC'.$

Hence C and C' fall together (why? Art. 56).   Q. E. D.

136. Theorem LXXV. — *The tangent-length from a vertex of a $\triangle$ to the in-circle equals half the perimeter of the $\triangle$ less the opposite side* (Fig. 93).

Proof. For the sum of $CE + CD + BD + BF$ is plainly $2a$ (why?); subtract this from the whole perimeter, $a + b + c$, and there remains $AE + AF = a + b + c - 2a$, or $AE = \dfrac{b + c - a}{2} = AF$. Q. E. D.

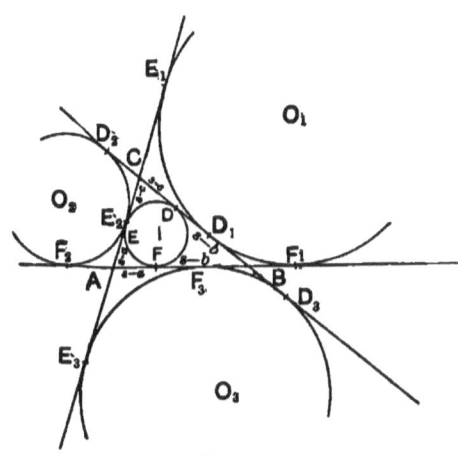

FIG. 93.

It is common and convenient to denote the **perimeter** (Fig. 93) (= measure round = sum of sides) by $2s$; then we see that the tangent-lengths from A, B, C, are $s-a$, $s-b$, $s-c$.

Corollary. The tangent-length from any vertex, A, of a $\triangle$ to the opposite ex-circle and the two adjacent ex-circles are s, $s-b$, $s-c$. Hence $s-a$, s, $s-b$, $s-c$, are the four tangent-lengths from any vertex, A, of a $\triangle$ to the in-circle and the three ex-circles.

These relations are useful and important.

137. Theorem LXXVI. — *There is a regular n-side.*

Proof. For the angle is a continuous magnitude (why?); hence there are angles of all sizes from zero to a round angle; hence there is an angle, the $\frac{1}{n}$ part of a round angle, such that, taken n times in addition, the sum will be a round angle. *Suppose* such an angle drawn, *whether or not we can actually draw it*, and suppose n such angles placed consecutively around any point O, so as to make a round angle. In other words, suppose n half-rays drawn cutting the round angle about O into n equal angles. Draw a circle about O, with (Fig. 94) any radius, and draw the n chords

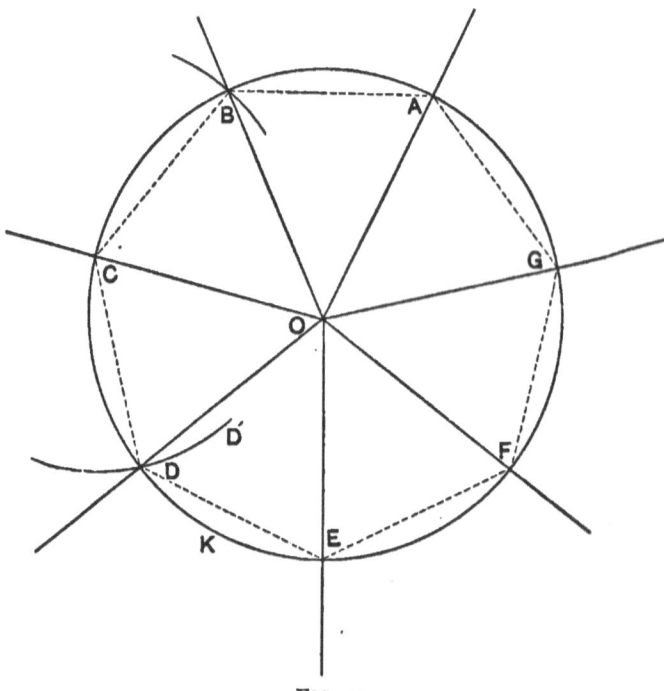

FIG. 94.

subtending the n equal central angles. These chords are all equal (why?), and subtend equal arcs, and they form an n-side. Moreover, the angle between two consecutive sides is constant in size, because it stands on the $\frac{n-2}{n}$ part of the circle. Hence the n-side is both equilateral and equiangular; that is, it is regular. Q. E. D.

Corollary. The inner angle of a regular n-side is the $\left(\frac{n-2}{n}\right)$ part of a straight angle.

Find the value in degrees of the inner angles of the first ten regular n-sides.

N.B. The foregoing demonstration merely settles the question of the existence or *logical possibility* of the regular n-side. The problem of *actually drawing* such a figure is one of the most intricate in all mathematics, and has been solved only for certain very special classes of values of n. But in order to discover the properties of the figure, it is by no means necessary to be able to draw it accurately. It is only since 1864 that we have known how to draw a straight line or ray exactly.

137 a.- Theorem LXXVII. — *The vertices of a regular n-side are encyclic* (Fig. 94).

Proof. Through any three vertices, as A, B, C, of a regular n-side, draw a circle K; about C with radius CB draw another circle. The fourth vertex D must lie on this circle (why?). If it lie on the circle K, then the angle $BCD =$ angle ABC, as is the case in the regular n-side. Neither can it lie off of K, as at D' or D'', because then the angle BCD' or BCD'' would not equal angle BCD (why?), and hence would not equal angle ABC. Hence the next vertex must lie on the same circle K, and so on all around. Q. E. D.

138. Theorem LXXVIII. — *The sides of a regular n-side are pericyclic* (that is, they all touch a circle).

118   GEOMETRY.   [TH. LXXIX.

Proof. For, on drawing the radii of the circumcircle K (Fig. 95) to the vertices, we get n congruent symmetric △

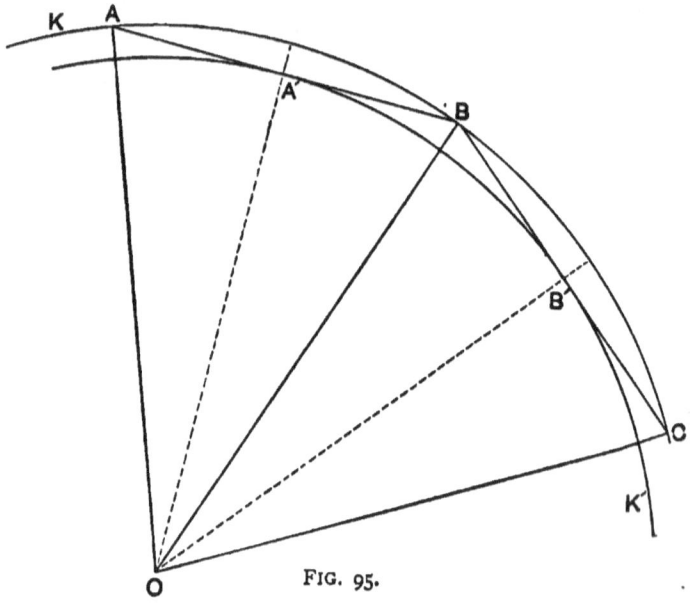

FIG. 95.

(why?). The altitudes of all are the same (why?); with this common altitude as radius draw another circle, K', about the same centre. It will touch each of the sides (why?). Q. E. D.

Corollary. The points of touch of the sides of the regular circumscribed n-side are mid-points of the sides.

139. Theorem LXXIX. — *The points of touch of a regular circumscribed n-side are the vertices of a regular inscribed n-side.*

Proof. Connect the points of touch consecutively. Then the △ so formed are all congruent (why?); hence the joining chords are equal; hence the arcs are equal; hence the Theorem. Q. E. D.

THE CIRCLE AS ENVELOPE.

*140a. Thus far we have regarded the circle from various points of view; from the most familiar it was seen to be the locus of a point in a plane at a fixed distance from a fixed point. An almost equally important conception of the curve treats it not as the *locus* of a point, but as the **envelope** *of a ray*. If the point P moves in the plane always equidistant from O, then its locus is the circle, on which it may always be found; also, if the ray R moves about in the plane always equidistant from O, then its *envelope* is the circle, on which it may always be found, on which it lies, which it continually touches. The point traces the circle, the ray envelops the circle, which is accordingly called the *envelope* (*i.e.* the *enveloped* curve — French *enveloppée*) of the ray. In higher mathematics the notion of the ray, instead of the point, as the determining element in the nature of a curve, attains more and more significance. In this text we are confined to the circle — the envelope of a ray in a plane, at a fixed distance from a fixed point.

*140b. It is not only rays, however, that may envelop a curve; but circles, and in fact any other curves. Thus, let the student draw a system of equal circles, having their centres on another circle; the envelope will at once be seen to be a pair of concentric circles. Let him also find the envelope of a system of circles equal and with centres on a given ray. In general, let him find the envelope of a circle whose centre moves on any given curve. Lastly, let him draw a large number of circles all of which pass through a fixed point, while their centres all lie on a fixed circle, and let him observe what curve they shadow forth as envelope.

Show that as the pole of a chord (or ray) traces a circle,

the chord itself envelops a concentric circle, and conversely.

Show that tangents from two points on a centre ray form a kite, and conversely. Also the chords of contact are parallel, and conversely.

O is the centre of a circle, P any point without it. Show how to find the point of touch of the tangents from P, by drawing a circle about O through P and a tangent where OP cuts the given circle.

CONSTRUCTIONS.

140. Hitherto, in our reasoning about concepts, figures have not been at all necessary, though exceedingly useful in making sharp and precise our imagination of the relations under consideration, in furnishing sensible examples of the highly general notions that we dealt with. The conclusions reached thus far all lie wrapt up in axioms and in our definitions of point, ray, and circle, and our work has been one of explication only; we have merely brought them forth to light. Our demonstrations have not presumed ability to draw accurately, and would remain unshaken if we could not draw at all. Nevertheless, for many practical purposes, it is extremely important and even indispensable that we actually make the constructions and draw the figures that thus far we have merely supposed made and drawn.

141. What is meant by drawing a ray, circle, or any line? Any mark, whether of ink or chalk, though a solid, may be treated as a line by abstraction. Only its length, not its width nor thickness, concerns us. How to make not just any mark, but some particular mark called for, is our **problem** ($\pi\rho o\beta\lambda\eta\mu a =$ anything thrown forward as a task), and

CONSTRUCTIONS.

its solution consists accordingly of two parts, the logical and the mechanical. The first is accomplished by fixing exactly in thought the position of all the geometric elements (points, rays, circles) in question; the second, by making marks that by abstraction may be treated as these elements. Now, a point is fixed as the join of two rays, a ray as the join of two points (by what axiom?); a circle is fixed or determined by its centre and radius (why?), or by three points on it (why?). Accordingly, when we know two rays through a point, or two points on a ray, or centre and radius, or three points of a circle, we know the point, or ray, or circle completely. The logical part of our work is finished, then, when we determine every point as the join of two known rays, every ray as the join of two known points, every circle as drawn through three known points or about a known centre with a known radius. The mechanical part of the solution requires us to put and keep a point in motion along a circle or a ray. Circular motion is brought about by the compasses already described (Art. 114), of which the shape is arbitrary, the necessary parts being merely a fixed point rigidly connected in any way with a movable point. But in the ruler one edge is supposed *made straight* to begin with, so that a pencil point gliding along it may trace a straight mark. Hence the use of the ruler is really illogical, since it *assumes* the problem of drawing a ray or straight line as already solved in constructing the straight edge. To say that, in order to draw a *straight line*, we must take a *straight edge* and pass a pencil point along it, is no better logically than to say that, in order to draw a circular line, we must take a circular edge and pass a pencil point along it. The question at once arises, How make the edge straight or circular in the first place? It was not until 1864 that Peaucellier won, though he did not at once receive, the Montyon prize from the French Academy by solving the thousand-year-old problem of imparting rectilinear motion to a point without guiding edge of any kind (Page ooo). But, though the ruler is logically valueless, it is practically invaluable, even after the great discovery of Peaucellier. Its edge being assumed as straight and of any desired length, and a pair of compasses of adjustable size being given, we now make the following **Postulates**:

I. *About any point may be drawn a circle of any radius.*

II. *Through any two points may be drawn a ray* (more strictly, a tract of any required length).

Corollary. On any ray from any point on it we may lay off a tract of any required length.

These are the only instruments used or postulates assumed in the constructions of Elementary Geometry.

142. The fundamental relations of rays to each other are two: Normality and Parallelism. Hence

Problem I. — *To draw a ray normal to a given ray.* Since there are many rays normal to a given ray, to make the problem definite we insert the limiting condition, **through a given point.** Two cases then arise:

A. *When the given point is on the given ray.* All we can do is to draw a circle about the point P. It cuts the ray at two points, A and A', symmetric as to P. Hence the midnormal of AA' is the normal sought. Hence any point on this normal lies on two circles of equal radius about A and A'. Hence (Fig. 96)

Solution. From the given point P lay off on the given ray two equal tracts PA, PA'. About A and A' draw two equal circles. Through their points of intersection draw their common chord. It is the normal sought.

Proof. For it is the mid-normal of AA', since it has two of its points equidistant from A and A', and P is the mid-point of AA'.

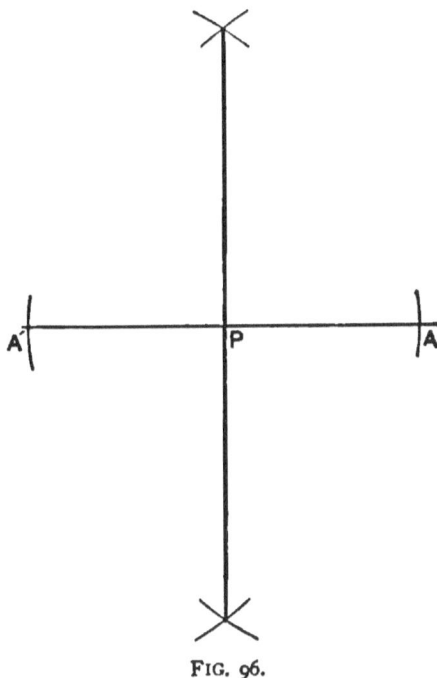

FIG. 96.

Query: What radius shall we take for the circles about A and A'?

B. *When the given point is not on the given ray.* All we can do is to draw a circle about the given point P. Let it cut the ray at A and A'. Then the mid-normal to AA' is the normal required (why?). Hence (Fig. 97)

Solution. Determine the points A, A' on the ray by a circle about the given point P; then proceed as in the first case (A).

124   GEOMETRY.

Proof. For the mid-normal of AA' goes through P (why?).

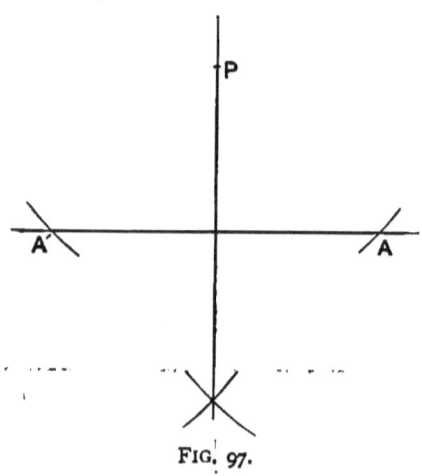

FIG. 97.

Query: What radius shall we take for the circle about P?

143. Problem II. — *To draw a parallel to a given ray.* Since there are many parallels to every ray, to make the problem definite we must insert the limiting condition, **through a given point**; then it becomes perfectly definite (why?). Manifestly the point must be not on the ray (why?). We now reflect that a transversal makes equal corresponding angles with parallels, and we have just learned to draw a normal transversal. Hence (Fig. 98)

Solution. Through the point draw a normal to the ray; through the same point draw a normal to this normal. It will be the parallel required.

Proof. For it goes through the point and is parallel to the ray (why?).

These two problems have been discussed at such length as being the hinges on which nearly all others turn. At the end of a problem is sometimes written Q. E. F. = *quod erat faciendum* = *which was to be done*, and translates the Euclidean ὅπερ ἔδει πρᾶξαι.

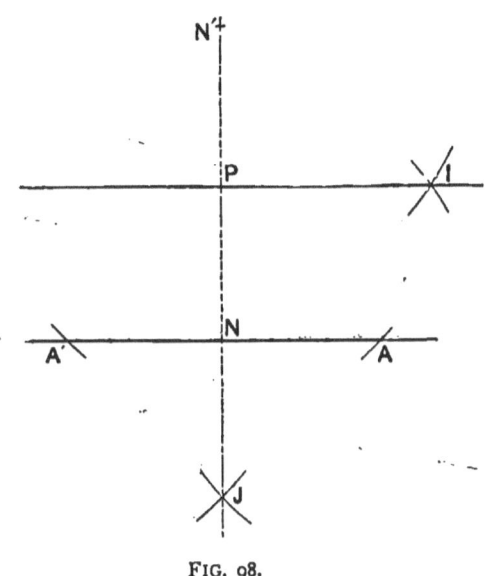

FIG. 98.

144. **Problem III.** — *To bisect a given tract*, or *to draw the mid-normal to a given tract*, AB.

Proceed as in Problem I.

145. **Problem IV.** — *To bisect a given angle.*

Solution. About the vertex draw any circle cutting the arms at A and A', and draw the mid-normal of AA'. It is the mid-ray sought (why?).

Corollary. Show how to bisect any circular arc AB.

146. **Problem V.** — *To bisect the angle between two rays whose join is not given* (Fig. 99).

We reflect that the join AA' of two corresponding points on the rays makes equal angles with the two rays that form the angle. Hence

Solution. From any point P of L draw the normal to it, cutting M at Q. From Q draw the normal to M. Bisect

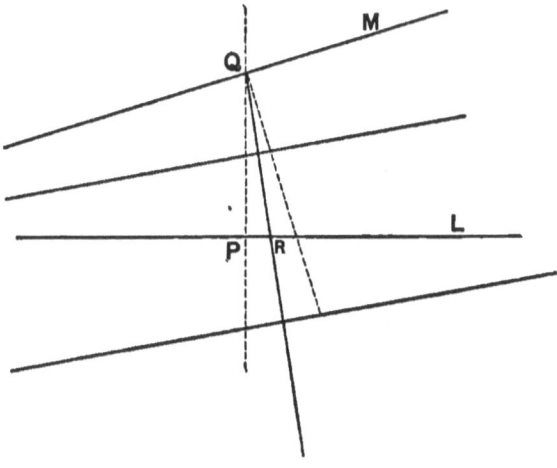

FIG. 99.

the angle at Q between these two normals by the mid-tract QR. Draw the mid-normal of QR. It is the mid-ray sought (why?).

147. Problem VI. — *To multisect a given tract AB* (Fig. 100).

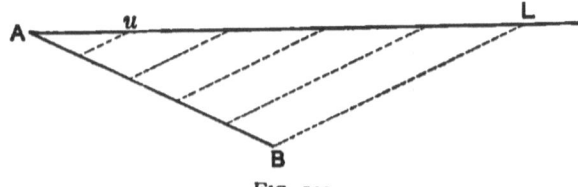

FIG. 100.

CONSTRUCTIONS. 127

Solution. Through either end of the tract, as A, draw any ray, and lay off on it from A successively n equal tracts, L being the end of the last. Draw BL. Through the ends of the equal tracts draw parallels to BL. They cut AB into n equal parts (why?).

148. **Problem VII.** — *To draw an angle of given size, i.e. equal to a given angle* (Fig. 101).

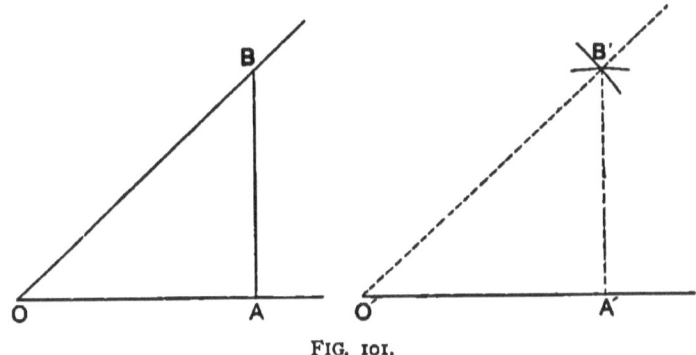

FIG. 101.

Solution. At any point A of either arm of the given angle O erect a normal to OA cutting the other arm at B. From any point O' on any other ray lay off $O'A' = OA$, and normal to the ray erect $A'B' = AB$ and draw $O'B'$. Then angle O' = angle O (why?).

When does this construction fail? How proceed then?

149. **Problem VIII.** — *To draw a tract of given length subtending a given angle and parallel to a given ray.*

Data: O the given angle, L the ray, a the length (Fig. 102).

Solution. Through any point P of either arm of the angle draw a parallel to the ray, and lay off on it towards the other arm a tract PA of the given length a. Through A draw a parallel to OP, cutting the other angle arm at Q; through Q draw a parallel to PA meeting OP at R. QR is the subtense sought (why?).

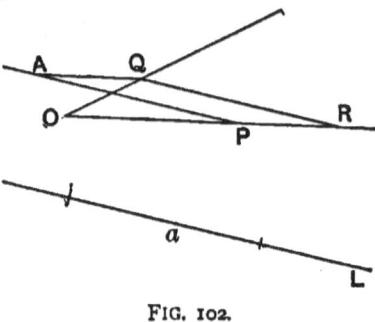

FIG. 102.

150. Problem IX. — To construct a $\triangle$:

A. *When alternate parts* (three sides or three angles) *are given*.

Solution. About the ends of one side AB, with the other sides for radii, draw circles meeting in C. Then ABC is the $\triangle$ sought (why?) (Fig. 103).

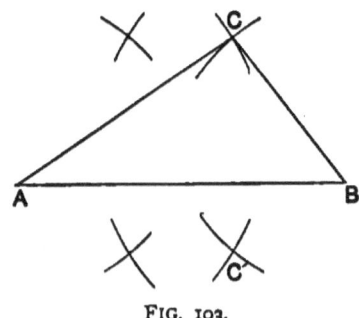

FIG. 103.

How many such $\triangle$ may be drawn on the same base AB? How are they related? When is the solution impossible?

When the angles are given, apply the construction of Problem VII. How many solutions are possible? What kind of △?

B. *When three consecutive parts* (two sides and included angle or two angles and included side) *are given*.

Solution. Apply the construction in Problem VII.

C. *When opposite parts* (two angles and an opposite side or two sides and an opposite angle) *are given*.

Solution. Apply the construction in Problem VII. When is the construction ambiguous?

D. *When two sides and the altitude to the third side are given*.

Solution. Through one end of the altitude draw a normal to it for the base; about the other end C as centre, with the sides as radii, draw circles cutting the base at A and A', B and B'; then ACB or $A'CB'$ is the △ required. Why?

E. *When two sides and the medial of the third side are given*.

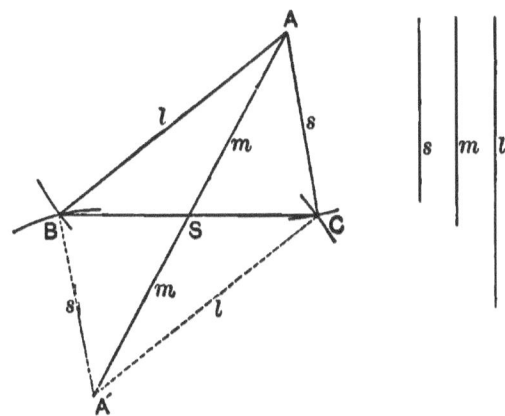

FIG. 104.

GEOMETRY.

If SA be the medial of BC, and SA' be symmetric with (Fig. 104) SA as to S, then $ABA'C$ is a parallelogram (why?); hence

Solution. Take a tract the double of the medial. About its ends as centres with the sides as radii draw circles and then complete the construction. How many △ fulfilling the conditions are possible? How are they related?

F. *When the three medials are given* (Fig. 105).

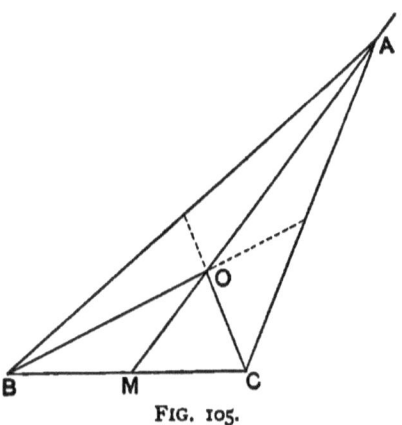

FIG. 105.

Solution. Remember that the medials trisect each other; construct the △ OBC according to (E), and draw OA counter to OM and twice as long.

151. Problem X. — *To construct an angle of given size and subtended by a given tract.*

Data: O *the given angle, AB the given tract* (Fig. 106).

Solution. Construct the angle BAD of given size (how?), draw the mid-normal of AB, meeting AD at P; also the normal to AD at A, meeting the mid-normal at S. About

CONSTRUCTIONS. 131

S as a centre with radius SA draw a circle; it touches AD at A (why?). The vertex V of the required angle may be anywhere on the arc AVB or on its symmetric $AV'B$ (why?).

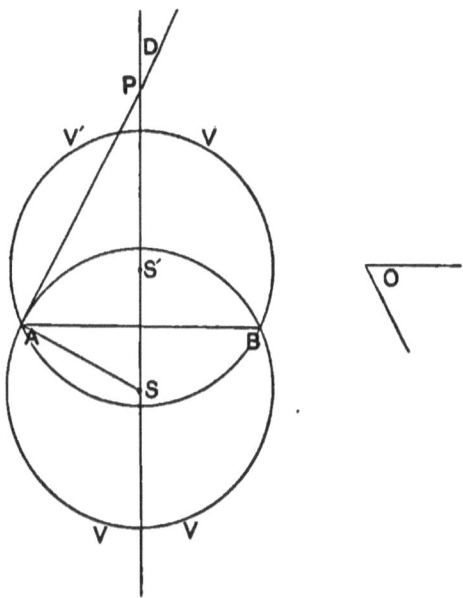

FIG. 106.

152. **Problem XI.** — *To draw a circle tangent to a given ray.*

Solution. About any point S with a radius equal to the distance of S from the ray, L, draw a circle; it will be a circle required (why?). If the circle must touch the ray L at a given point P, then S must be taken on the normal to L through P (why?). If, besides, the circle must go through a given point Q, then S must also be on the mid-normal of PQ (why?). Hence the construction.

153. Problem XII. — *To draw a circle touching two given rays.*

The centre may be anywhere on either mid-ray (why?). If now the circle is to touch a **third given ray**, the centre must be also on another mid-ray; that is, it must be the intersection of two mid-rays of the three angles of the three rays. There are four such intersections — what are they? Complete the construction. See Fig. 60.

154. Problem XIII. — *To draw a circle through two points.*

The centre S may be anywhere on the mid-normal of the tract AB between the points (why?), the radius is — what? If now the circle is to pass **through a third point** C, then S must also be on the mid-normal of BC and CA (why?). There is one, and only one, such point (why?); complete the construction. When is the construction impossible?

155. Problem XIV. — *To draw a circle through two given points and tangent to a given ray; or, tangent to two given rays and through a given point.*

This double problem is mentioned here because it must naturally present itself to the mind of the student; but the solution involves deeper relations than we have yet explored. See Art. ooo.

Several of the foregoing problems were indefinite, admitting any number of solutions: these latter taken all together form a **system** or *family*. Problems concerning parallelograms and other 4-sides may often be solved on cutting the 4-side into two △.

156. Problem XV. — *To inscribe a regular 4-side* (square) *in a circle* (Fig. 107).

CONSTRUCTIONS. 133

Solution. Join consecutively the ends of two conjugate diameters. The 4-side formed is inscribed (why?) and is a square (why?).

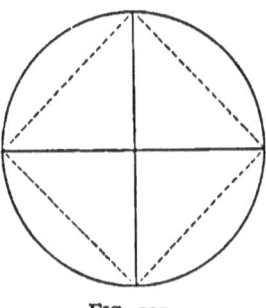

FIG. 107.

157. **Problem XVI.** — *To inscribe a regular 6-side in a circle.*

Solution. Apply the radius six times consecutively as a chord to the circle (Fig. 108). The figure formed will be the regular 6-side (why?).

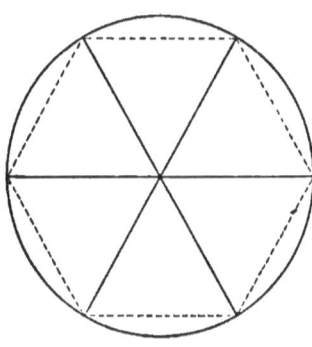

FIG. 108.

N.B. This seems to have been one of the first geometric problems ever solved. The Babylonians discovered that six radii thus applied would compass the circle, and having already divided the circle into 360 steps, they accordingly divided this number by 6 and thus obtained 60 as the basis of the famous *sexagesimal* notation, which long dominated mathematics and still maintains its authority undiminished in astronomy and chronometry.

In more difficult problems it is often advisable, or in fact necessary, to **suppose the problem solved**, the construction made, and investigate the relations thus brought to light. Then the facts thus discovered may be used regressively in making the construction required. This method is illustrated in the following:

158. Problem XVII. — *To draw a square with each of its sides through a given point.*

Let A, B, C, D, be the four given points, and suppose (Fig. 109) $PQRS$ to be the square properly drawn. Draw

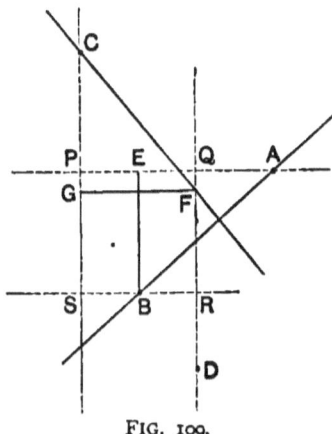

FIG. 109.

AB, cutting a side of the square, and through B draw BE parallel to the side cut. Through a third point C draw a normal to AB, meeting QR in F. Also draw FG parallel to PQ. Then the △ ABE and CFG are congruent (why?).

Hence we discover that $CF = AB$. This fact is the key to the

Solution. Join two points A and B; from a third, C, lay off CF equal and normal to AB. The join of D, the fourth point, and F is one side of the square in position (why?). Let the student complete the construction and show that four squares are possible.

159. Problem XVIII. — *To trisect a given angle.*

Suppose the problem solved and the ray OT to make $\angle TOB = \tfrac{1}{2} TOA$ (Fig. 110).

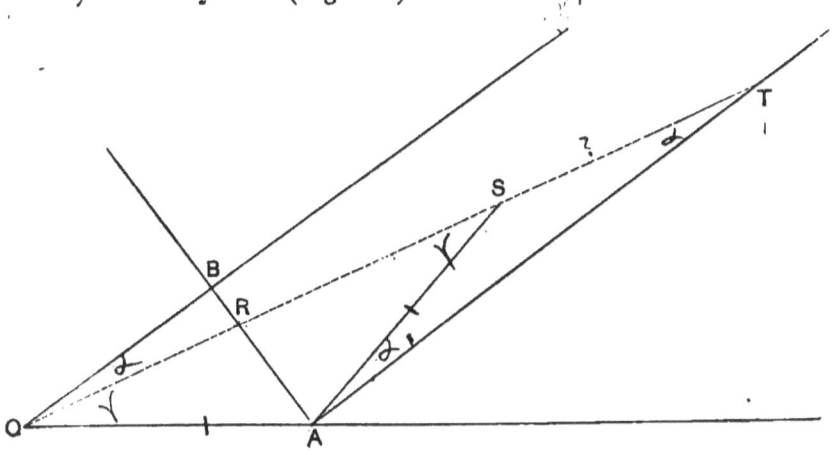

FIG. 110.

From any point A of the one end of the angle draw a parallel and a normal to the other end; also draw to the trisector a tract $AS = OA$. Then the following relations are evident:

$\angle AOS = \angle ASO = \angle SAT + \angle STA$;
but $\angle AOS = 2 \angle TOB = 2\, STA$;
hence $\angle STA = \angle SAT$, and $ST = SA$.

Again, $\angle SAR = \angle SRA$, being complements of equal angles; hence $SA = SR$, $TR = 2\,OA$. Hence

Solution. From any point A of either arm of the given angle draw a parallel and a normal to the other arm; then, with one point of a straight-edge fixed at the vertex O, *turn the edge until the intercept between the normal and the parallel equals* $2\,OA$. But to do this we need a *graduated edge, or a sliding length $2\,OA$ on the edge itself*. Accordingly, this construction, while simple, useful, and interesting, is not *elementary geometric* in the sense already defined. To discover such a solution for this famous problem, has up to this time baffled the utmost efforts of mathematicians.

EXERCISES II.

1. State and prove the reciprocals of Exercises 9, 10, 11, page 71.

2. Find a point on a given ray, the sum of whose distances from two fixed points is a minimum.

3. The same as the foregoing, difference supplacing sum.

4. A and A', B and B', C and C', are symmetric as to MN. Show that $\triangle ABC \equiv \triangle A'B'C'$.

5. The inner and outer mid-rays of the basal angles of a symmetric $\triangle$ form a kite.

6. The inner mid-rays of the angles of a trapezium form a kite with two right angles.

7. The joins, of the mid-points of the parallel sides of an anti-parallelogram, with the opposite vertices, form a kite.

8. The mid-rays of the angles at the ends of the transverse axis of a kite cut the sides in the vertices of an anti-parallelogram.

EXERCISES II.

9. How must a billiard ball be struck so as to rebound from the four sides of a table and return through its original place?

10. Trace a ray of light from a focus P, to another given point Q, reflected from a convex polygonal mirror.

11. A ray of light falls on a mirror M, is reflected along S to a second mirror M', is thence reflected along T. Remembering that the angle of incidence equals the angle of reflection, show that the angle between the original ray R and its last reflection T is twice the angle between the mirrors (angle $RT = 2$ angle MM'). On this theorem is grounded the use of the **sextant**.

12. Two mirrors stand on a plane and form an inner angle of 60°; a luminous point P is on the mid-ray of this angle (or anywhere within it); how many images of P are formed? How are they placed? What if the angle of the mirrors be $1/n$ of a round angle?

This theorem is beautifully illustrated in the **kaleidoscope**.

13. A regular n-side has n axes of symmetry concurring in the centre of the n-side, which centre is equidistant from the sides of the n-side.

14. How do these axes lie when n is even? when n is odd? Show that if n be even, the centre is a centre of symmetry.

15. The half-rays from centre to vertices of a regular n-side form a regular pencil of n half-rays, and those from the centre normal to the sides, another regular pencil; also the half-rays of each pencil bisect the angles of the other.

16. In a figure with two rectangular axes of symmetry each point, with three others, determines a rectangle, and each ray, with three others, a rhombus.

17. Find the axes of symmetry of two given tracts.

18. A regular $\triangle$, along with the figure symmetric with it as to its centre, determines a regular 6-angle (6-pointed star).

19. Two congruent squares, the diagonals of one lying on the mid-parallels of the other, form a regular 8-angle; also find the lengths of the intercepts at the corners.

20. The outer angle of a regular n-side is m times the outer angle of a regular mn-side.

EXERCISES III.

1. A circle with its centre on the mid-ray of an angle makes equal intercepts on its arms.

2. Tangents parallel to a chord bisect the subtended arcs, and conversely.

3. Tangents at the end of a diameter are parallel.

4. A and B are ends of a diameter, C and D any other two points of a circle; E is on the diameter, and angle $AED = 2$ angle CAD; find the possible positions of E.

5. From n points are drawn $2n$ equal tangent-lengths to a circle; where do the points lie?

6. In a circumscribed hexagon, or any circumscribed $2n$-side, the sums of the alternate sides are equal.

7. If the vertices of a circumscribed quadrangle, hexagon, or any $2n$-side, be joined with the centre of the circle, the sums of the alternate central angles will be equal.

8. The sums of the alternate angles of an encyclic $2n$-side are equal, namely, each sum is $(n-1)$ straight angles.

9. The joins of the ends of two parallel chords are symmetric as to the conjugate diameter of the chords.

EXERCISES III.

10. A centre ray is cut by two parallel tangents. Show that the intercepts between tangent and circle are equal.

11. Normals to a chord from the ends of a diameter make, with the circle, equal intercepts on the chord.

12. The joins of the ends of two diameters are parallel in pairs, and form a rectangle, and meet any two parallel tangents in points symmetric in pairs as to the centre.

13. The joins of the ends of two parallel chords meet the tangents normal to the chords in points whose other joins are parallel to the chords.

14. A chord AB is prolonged to C, making $BC =$ radius, and the centre ray CD is drawn; show that one intercepted arc is thrice the other.

15. The intercepts, on a secant, of two concentric circles are equal.

16. A chord through the point of touch of two tangent circles subtends equal central angles in the circles.

17. Two rays through the point of touch of two tangent circles intercept arcs in the circles whose chords are parallel.

18. The transverse joins of the ends of parallel diameters in two tangent circles go through the point of tangence.

19. Four circles touch each other outerly in pairs: 1st and 2d, 2d and 3d, 3d and 4th, 4th and 1st; show that the points of touch are encyclic.

20. Show that three circles drawn on three diameters OA, OB, OC intersect on the sides of the $\triangle ABC$.

21. Find the shortest and the longest chord through a point within a circle.

22. In a convex 4-side the sum of the diagonals is greater than the sum of two opposite sides, less than the sum of all the sides, and greater than half the sum of the sides.

140   GEOMETRY.

23. Three half-rays trisect the round angle O; on each is taken any point, as A, B, C. Find a point M such that the sum $MA + MB + MC$ is the least possible (a minimum).

24. Two tangents to a circle meet at a point distant twice the radius, from the centre; what angles do they form?

25. The intercept of two circles on a ray through one of their common points subtends a constant angle at the other.

26. What is the envelope of equal chords of a circle?

27. Two movable tangents to a circle intersect under constant angles; find the envelope of the mid-rays of these angles.

28. The vertex V of a revolving right angle is fixed midway between two parallels, and its arms cut the parallels at A and B; find the envelope of AB.

29. From a fixed point P a normal PN is drawn to a movable tangent T of a circle, and through the mid-point M of PN there is drawn a parallel to T; find its envelope.

30. The vertices of a $\triangle$ are V_1, V_2, V_3; the mid-points of its sides are M_1, M_2, M_3; the feet of its altitudes are A_1, A_2, A_3; the inner bisectors of its angles meet the opposite sides at B_1, B_2, B_3; and the outer bisectors at B'_1, B'_2, B'_3; its centroid is C, its in-centre is I, its circum-centre is S; its angles are α_1, α_2, α_3, and their complements are α'_1, α'_2, α'_3. Express through these six angles the angles between: (1) V_1A_1 and V_1V_2; (2) A_1A_2 and V_2V_3; (3) A_1A_2 and V_1A_1; (4) A_1A_2 and A_2A_3; (5) M_1A_2 and V_3V_1; (6) M_1A_2 and V_2V_3; (7) M_1A_2 and M_1A_3; (8) A_1M_2 and A_1M_3; (9) A_1M_2 and A_1A_2; (10) SV_1 and SV_2; (11) SV_1 and V_1V_2; (12) SV_1 and V_2A_2; (13) IV_1 and IV_2; (14) IV_1 and V_2A_2; (15) VB'_1 and $V_2B'_2$.

31. Find the locus of the mid-points of chords through a fixed point upon, within, or without a fixed circle.

32. Find the locus of the mid-points of the intercepts of a secant between a fixed point and a fixed circle.

33. As the ends of a ruler slide along two grooves normal to each other, how does its mid-point move?

34. Two equal hoops move along grooves normal to each other and touch each other; how does the point of touch move?

35. Orthocentre O, centroid C, circum-centre S, and centre F of Feuerbach's (9-point) circle, of a $\triangle$ are collinear, and $OC = 2\,CS$ (Euler), $OF = FS$.

36. Two parallel tangents meet two diameters of a circle at the vertices of a parallelogram concentric with the circle.

37. The inner mid-rays of the angles of a 4-side form an encyclic 4-side.

38. The outer mid-rays of the angles of a 4-side form an encyclic 4-side. How are the 4-sides of 37 and 38 related?

39. The circum-centres of the four $\triangle$ into which a 4-side is cut by its diagonals are the vertices of a parallelogram.

40. The circum-centres of the two pairs of $\triangle$, into which a 4-side is cut by its diagonals in turn, are how related to each other and to the centres in 39?

EXERCISES IV.

1. Construct a square, knowing

   (*a*) Its side; or (*b*) its diagonal.

2. Construct a rectangle, knowing

   (*a*) Two sides; or (*b*) a side and a diagonal; or (*c*) either a side or a diagonal and the angle of either with the other; or (*d*) a diagonal and its angles with the other diagonal.

3. Construct a parallelogram, knowing

(*a*) Two sides and one angle; or (*b*) a side, a diagonal, and the included angle; or (*c*) two sides and the opposite diagonal; or (*d*) two sides and the included diagonal; or (*e*) two diagonals and a side; or (*f*) two diagonals and their angles with each other.

4. Construct an anti-parallelogram, knowing

(*a*) Its parallel sides and the distance between them; (*b*) two adjacent sides and their included angle; (*c*) two adjacent sides and the angle between the non-parallel sides; (*d*) a diagonal and two adjacent sides; (*e*) a diagonal, a side, and the included angle.

5. Construct a kite, knowing

(*a*) Two sides and an axis; (*b*) two sides and the included angle; (*c*) a side and the axes.

6. Construct the rays equidistant from three given points.

7. Draw a ray through a given point equally sloped to two given rays.

8. A square has one vertex at a given point, and two others on two given parallel rays; draw it.

9. Hypotenuse and sum of sides of a right $\triangle$ are given; draw it.

10. Construct a regular 2^n-side, and a regular 3.2^n-side.

11. Find the centre of a given circular arc.

12. Trisect a right angle.

13. Two points, A and B, of a ray are given; find any number of points of the ray without drawing it, and without opening the compasses more than AB.

14. Find a point on a given ray or given circle that has a given tangent-length with respect to a given circle.

15. Through a given point draw a secant on which a given circle shall make a given intercept.

16. **Draw four common tangents to two given circles.**

17. Draw a ray touching a given circle and equidistant from two given points.

18. Draw a ray on which two given circles shall make two given intercepts.

19. With three given radii draw three circles, each touching the other two.

20. Draw a circle touching the radii and the arc of a given sector.

21. Draw a circle touching two given equal intersecting circles and their centre ray.

22. On the central intercept of two equal intersecting circles as diameter draw a circle; then draw a circle touching the three circles.

23. Three equal circles touch each other outerly; draw a circle touching the three.

24. Find a point from which two given apposed tracts appear to be equal.

25. Through two given points of a given circle draw a circle that shall cut a third circle orthogonally.

26. Construct a △, knowing

(*a*) The feet of the altitudes; (*b*) the foot of one altitude and the mid-points of the other two sides; (*c*) the three ex-centres; (*d*) two ex-centres and the in-centre.

27. Draw through a given point a ray that shall form with the sides of a given angle a △ of given perimeter. Hint: Use the properties of ex-circles.

28. Draw a 5-side, knowing the mid-points of the sides.

29. On a tract AB there is drawn a regular 3-side; draw on it a regular 6-side. Generalize the problem, changing 3 into n, 6 into $2n$, and solve it.

30. Given a regular n-side; draw a regular $2n$-side having the original n vertices for alternate vertices. Do not use the circumcircle.

AREAL RELATIONS.

Geometrica geometrice.

160. Hitherto our attention has been fastened exclusively on points and lines as composing figures. We have regarded no higher extents than those of one dimension. The surface, or extent of two dimensions, bounded by a circle or the sides of a triangle, we have not considered at all, but only the circle or triangle itself. Moreover, our comparisons as to size have referred exclusively to magnitudes directly superposable and homœoidal both in themselves and between themselves, such as tracts, angles, arcs of the same circle. Such comparisons are not suited to set forth the sharp distinction between congruence and equality, inasmuch as congruent tracts, angles, arcs are equal, and equal tracts, angles, arcs are congruent. But, as we now advance to the discussion of two-dimensional extents, the distinction in question becomes essential and regulative. Accordingly we premise the following:

Def. A surface or two-dimensional extent considered solely as to its size, or the amount of a two-dimensional extent, is called an **Area**.

N.B. Where no confusion will result, we shall designate the area bounded by a border by the name of the border itself: thus, circle for the area bounded by a circle, etc.

161. As we shall have frequent occasion to compound two areas into one and to divide one area into two, in order to treat these processes logically we must first define them precisely.

Def. If any two points of the border of a surface bounded by a single continuous line be joined, the surface is said to be cut or divided into two parts, and the section-line counts as part-border both of the one part and of the other. Thus SL is such a section-line (Fig. 111).

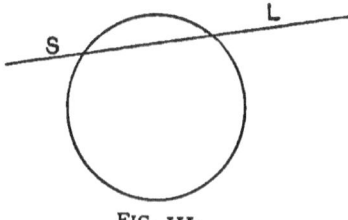

FIG. 111.

N.B. It is essential that the border be single and continuous; that is, that the surface be *simply compendent*. If the surface be doubly compendent, as a ring, then the section-line may or may not cut it into two parts. This doctrine of the compendency of surfaces is a creation of Riemann's, with which we have at present no further concern (Fig. 112).

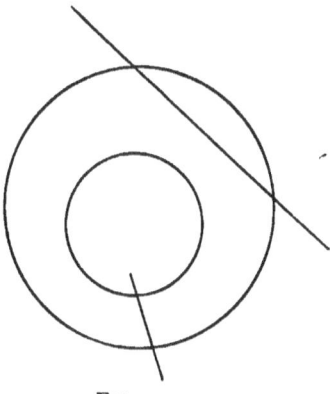

FIG. 112.

Any part of the border of a surface may be treated as the beginning or the end of the surface, and all the rest of the border as the end or the beginning.

162. When two areas, or surfaces, are apposed, the beginning of the second fitting on the end of the first, the two congruent part-borders herewith ceasing to be any part of the border of either, the whole area bounded by the remaining part-borders of the two, namely, the beginning of the first and the end of·the second, is called the **sum** of the two areas thus apposed ; *e.g.* we may appose two equal semi-circles and get a circle as the sum ; or two congruent △ and

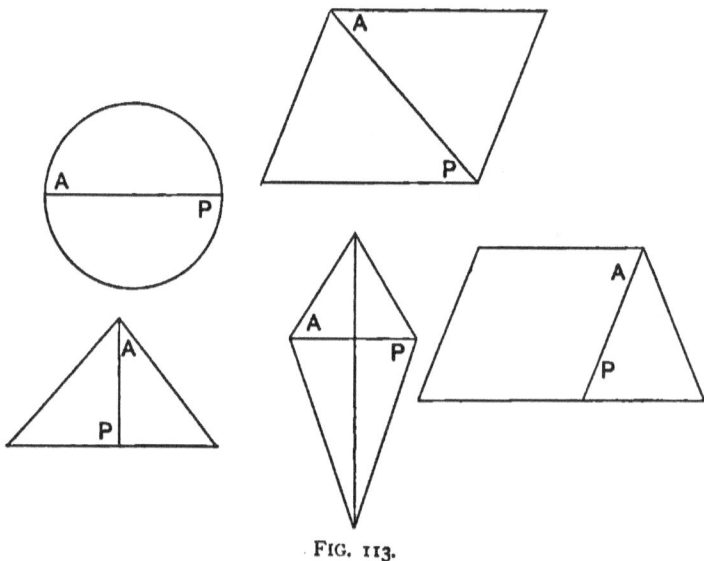

FIG. 113.

get a parallelogram as the sum ; or two congruent right △ and get a symmetric △ as sum ; or two symmetric △ and get a kite as sum ; or a parallelogram and a symmetric △ and get an anti-parallelogram as sum (Fig. 113).

Def. The areas apposed are called *parts* or **summands** of the **sum.**

CRITERIA OF EQUALITY.

163. 1. Two surfaces not congruent but divisible into the same number of parts so that for each part of either there is a congruent part of the other, are said to be *equal in area*, to *agree in area*, to have *equal areas*, or simply to be *equal*. These four phrases may be used indifferently, according to convenience.

2. Two surfaces not congruent, but which may be made congruent by addition, subtraction, or both, in case of each, of the same surfaces, or surfaces congruent in pairs, are said to be *equal in area*.

3. Two surfaces, each equal by either of the foregoing tests, to the same third surface, or to one of two equal surfaces, are themselves said to be *equal*.

These *three criteria* will serve our present purposes. A most important fourth criterion will be introduced at the proper place.

It thus appears that while congruence implies sameness as to both *size* and *shape*, equality implies sameness as to *size* only.

164. When the borders of two plane surfaces (the only ones that we deal with) are congruent, the surfaces are themselves congruent and their areas are equal. This follows from the homœoidality of the plane. For let A and A', B and B', be two pairs of corresponding points in the borders; then the ray AB will fit on the ray AB, point for point throughout; and so for any other pair of corresponding rays. Thus the one surface fits point for point, line for line, on the other.

165. The conditions of congruence among △ are known. The most important condition of congruence between two parallelograms is this:

Theorem LXXX. — *Two parallelograms with the sides of the one equal respectively to the sides of the other, and the angles of the one either equal or supplemental respectively to the angles of the other, are congruent* (Fig. 114).

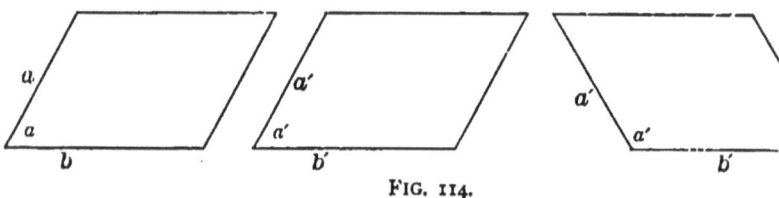

FIG. 114.

For, if $a = a'$, $b = b'$, and $\alpha = \alpha'$, then plainly we may fit the one parallelogram perfectly on the other. But if $\alpha =$ supplement of α', then $\alpha = \beta'$, and we may again fit the parallelograms. Q. E. D.

Corollary. Two rectangles having a pair of adjacent sides of one equal to a pair of adjacent sides of the other are congruent.

166. *Def.* Any tract forming part of the border of a closed figure and treated as its lowest part may be called its **base**.

Def. The greatest normal distance from a point of the figure to this base ray is called the **altitude** or height of the figure (Fig. 115).

In the diagrams *a* denotes altitude and *b* base — a convenient notation, which will be generally employed.

The notions of base and altitude are correlate, implying

each other, but they are *not* in general interchangeable ; the relation between them is not a mutual or reciprocal relation.

Def. The base and altitude of an areal figure may be called its *two dimensions*.

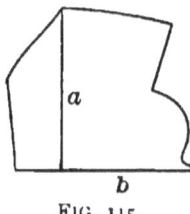

FIG. 115.

167. In only one figure are these dimensions interchangeable, namely, in the *rectangle*, and it is this fact that recommends the rectangle as the standard figure in comparison of areas.

A rectangle whose two adjacent sides, or whose altitude and base, or whose two dimensions, are a and b, will be denoted by the symbol *rect. ab*, or ▭ *ab*, or simply *ab*, rectangle being always understood.

If $a = AB$ and $b = BC$, we may write $AB \cdot BC$ and read *rectangle of AB* and *BC*, the dot (·) being used *not* to denote multiplication but merely to separate and distinguish base and altitude.

168. **Theorem LXXXI.** — *The sum of two rectangles that agree in one dimension is a third rectangle agreeing with each in that dimension and having for its second dimension the sum of the second dimensions of the summands.*

Data: $ABCD$ and $EFGH$ two rectangles agreeing in one dimension, $BC = EH$.

Proof. Appose BC and EH ; then AF becomes a single tract (why?), namely, the sum of the tracts AB and EF.

Also DG becomes similarly a single tract equal to AF. Hence $AFGD$ is a rectangle (why?), and by definition it is the sum of the two component rectangles; moreover, it

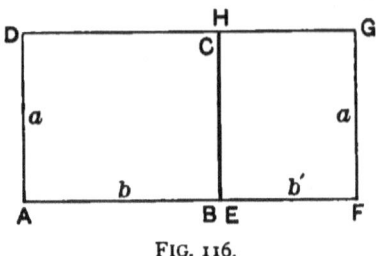
FIG. 116.

agrees with each in one dimension, while its second dimension, AF, is the sum of the two second dimensions, AB and EF. Q. E. D.

Corollary. Symbolically
$$ab + ab' = a(b + b'),$$
$$ab + a'b = (a + a')b,$$
$$aa' + ab' + ba' + bb' = a(a'+b') + b(a'+b') = (a+b)(a'+b').$$
Draw a figure exhibiting this last relation.

169. Theorem LXXXII. — *Two rectangles that agree in both dimensions are equal.*

For plainly they are congruent. Q. E. D.

170. Theorem LXXXIII. — *Two rectangles agreeing in one dimension but not in the other are unequal.*

Data: $ABCD$ and $EFGH$ two rectangles agreeing in one dimension, $AB = EF$, but not in the other, $BC \neq EH$.

Proof. Fit AB on EF; then DC will fall above or below HG according as $BC > EH$ or $BC < EH$; the whole of one rectangle fits on part of the other, *i.e.* they are unequal.   Q. E. D.

Corollary.   The rectangle with the greater dimension is the greater.

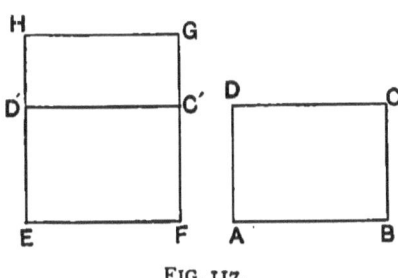

FIG. 117.

171. Theorem LXXXIV. — *Two rectangles agreeing in area and in one dimension agree in the other also.*

Proof. For if unequal in the second dimension they would be unequal in area (why?) ; but they are equal in area; hence, etc.   Q. E. D.

N.B.   In symbols,

if $ab = cd$ and $a = c$, then $b = d$.

172. Theorem LXXXV. — *A rectangle and a parallelogram that agree in both dimensions agree in area also.*

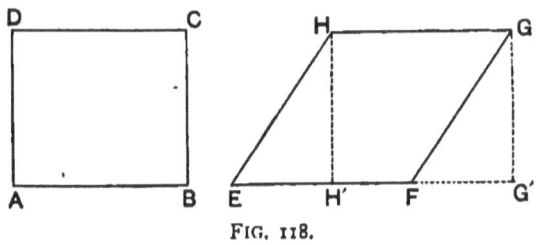

FIG. 118.

Data: A rectangle $ABCD$ and a parallelogram $EFGH$, having $AB = EF$ and $AD = HH'$.

Proof. Draw HH' and GG' normal to EF. Then the right $\triangle$ $EH'H$ and $FG'G$ are congruent (why?); take away the first from the trapezoid $EG'GH$, and there is left the rectangle $H'G'GH$; take away the second, and there is left the parallelogram $EFGH$; hence the parallelogram equals the rectangle. Moreover, this latter is congruent with the rectangle $ABCD$ (why?); hence the parallelogram $EFGH$ equals the rectangle $ABCD$. Q.E.D.

Corollary 1. Parallelograms agreeing in both dimensions are equal.

Corollary 2. Parallelograms agreeing in area and in one dimension agree in the other.

Corollary 3. Parallelograms agreeing in one but not in the other dimension are unequal.

Corollary 4. Parallelograms agreeing in one dimension but not in area do not agree in the other dimension.

Corollary 5. Parallelograms agreeing in area but not in one dimension do not agree in the other.

Corollary 6. Equal parallelograms with equal bases along the same ray, and lying on the same side of that ray, have the sides opposite the bases in the same ray.

173. Theorem LXXXVI.—*A $\triangle$ has half the area of a rectangle with the same dimensions.*

Proof. Complete the $\triangle$ ABC into the parallelogram $ABCD$; then ABC and DCB are two congruent $\triangle$ whose sum is the parallelogram $ABCD$ of the same dimensions; *i.e.* each is half the parallelogram; and this latter has the same area as the rectangle of the same dimensions; hence the $\triangle$ has half the area of the rectangle of the same dimensions. Q.E.D.

Scholium. We may express this fact by saying that the △ equals half the rectangle (or parallelogram) of its base and altitude.

Corollary 1. A △ is determined in area by its two dimensions, base and altitude.

Corollary 2. △ agreeing in dimensions agree in area.

Corollary 3. △ agreeing in area and in one dimension agree in the other also.

Corollary 4. △ with equal bases along the same ray and with vertices in either of two parallels equidistant from the base are equal.

Corollary 5. Equal △ with equal bases along the same ray have their vertices in two parallels equidistant from the base.

Corollary 6. A medial bisects the area of the △.

Corollary 7. In general, when any two of the three related magnitudes, base, altitude, and area of rectangle, parallelogram or △ are known, the third is also known univocally.

Corollary 8. So far as size is concerned, the dimensions in rectangle, parallelogram, and △ may be exchanged.

174. Since a △ is half the rectangle of the same dimensions, and since we can add rectangles agreeing in one dimension, we may also add △ agreeing in one dimension, preserving the same. Thus by apposing the bases b and b' of two (Fig. 119) △ agreeing in altitude a, we get half of a rectangle a $(b + b')$, equal to a △ with same altitude and base $b + b'$, the sum of the bases. By apposing two △ along the common base b we get half of the rectangle

$(a+a')\,b$ equal to a triangle with the common base b and with altitude the sum of the altitudes. Hence,

Theorem LXXXVII. — *The sum of two △ agreeing in one dimension equals a third △ with the same dimension, its second dimension being the sum of the second dimensions of the summands.*

Scholium. In symbols

$$ab + ab' = a\,(b+b'),\ ab + a'b = (a+a')\,b.$$

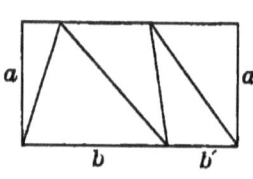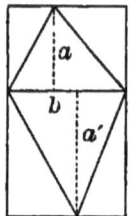

FIG. 119.

MISCELLANEOUS APPLICATIONS.

*175. Theorem LXXXVIII. — *When two △ lie apposed on the same base:* I. *That base bisects the join of the two vertices, if the △ are equal.*

Data: ABC and ABC' the equal apposed △ (Fig. 120).

Proof. The altitudes CD and $C'D'$ are equal when the △ are equal (why?); hence △ CDI and $C'D'I$ are congruent; hence $CI = C'I$. Q.E.D.

Conversely,

II. *The △ are equal, if that base bisects the join of the vertices.*

Proof. The △ CDI and $C'D'I$ are again congruent (why?); hence $CD = C'D'$, hence ABC and ABC' are equal. Q. E. D.

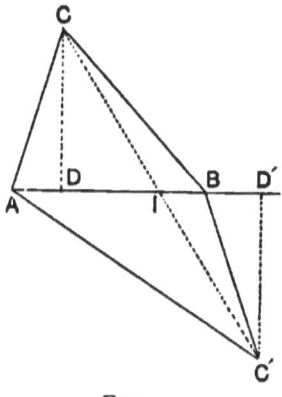

FIG. 120.

Def. Two parallels to the sides of a parallelogram through any point within it divide it into four parallelograms, and each pair of opposites we may call *complemental*.

*176. **Theorem LXXXIX.** — *When the common vertex of the complementals is on the diagonal, the pair on opposite sides of the diagonal are equal.*

Data: $ABCD$ the parallelogram, P the point on the diagonal AC, PD and PB the complementals (Fig. 121).

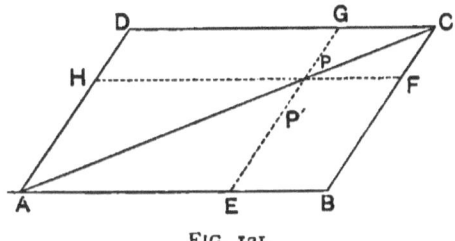

FIG. 121.

156   GEOMETRY.   [TH. XC.

Proof. From the equal △ ADC and ABC take away the pairs of equal △ APH and APE, CPG and CPF; there remain the equal complementals PD and PB.   Q. E. D.

Conversely, *When the complementals are equal, the common vertex is on the diagonal.*

Proof. If P moves off from the diagonal, as towards E, then PG will be lengthened and PE shortened without affecting either PH or PF; *i.e.* one of the equal complementals will be increased and the other decreased; hence, for P not on the diagonal the complementals are unequal; hence the theorem, by contraposition.   Q. E. D.

Corollary. Parallelogram $CH = $ parallelogram CE, and parallelogram $AF = $ parallelogram AG.

177. Theorem XC. — *A trapezoid equals half the rectangle of its altitude and the sum of its bases, or equals the rectangle of its altitude and the mid-parallel to its bases.*

Data: $ABCD$ a trapezoid, MP the mid-parallel (Fig. 122).

Proof. Draw $B'C'$ parallel to AD; hence, etc.   Q. E. D.

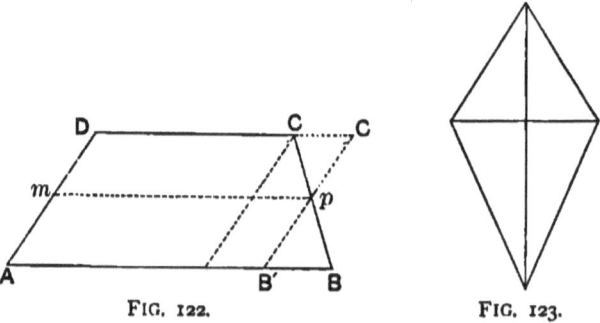

FIG. 122.   FIG. 123.

178. Theorem XCI. — *A kite equals half the rectangle of its diagonals* (Fig. 123).

Let the student deduce the proof from the figure.

SQUARES.

179. We have seen that the *rectangle* is distinguished among parallelograms by the interchangeability of its dimensions; among rectangles the **square** is distinguished by the equality of its dimensions. Hence it is determined completely by one dimension; hence we may reason about squares and compare them through their dimensions more readily than is possible with rectangles.

180. Theorem XCII. — *The square on the sum of two tracts equals the sum of the squares on the tracts and twice the rectangle of the tracts.*

Data: a and b two tracts, AB their sum, $ABCD$ the square on that sum (Fig. 124).

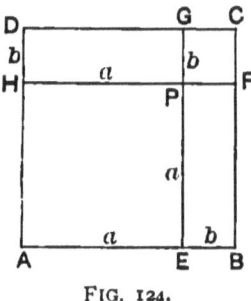

FIG. 124.

Proof. Draw EG parallel to BC and HF parallel to CD; they cut the whole square into four parts, namely, the square on a, the square on b, and the two congruent rectangles ab and ab. Q. E. D.

Corollary. The square on a tract equals four times the square on half the tract.

181. Theorem XCIII. — *The square on the difference of two tracts equals the sum of the squares on the tracts, less twice the rectangle of the tracts.*

Proof. In the preceding figure treat AB or $a+b$ as the one tract and EB or b as the other; then AE or a is the difference. The square AC is the square on the tract $a+b$; increase it by the square PC on b, so that this square is to be counted *twice* and thought as *doubly laid* in the figure; now strip off the rectangle HC or $(a+b)b$, and its congruent BG; there remains the square AP on the difference a. Q. E. D.

182. Theorem XCIV. — *The rectangle of the sum and difference of two tracts equals the difference of the squares on those tracts.*

Proof. In the same figure treat $a+b$ as the one tract and a as the other, so that $2a+b$ is the sum and b the difference of the tracts. Then the two rectangles EC and HG agree in one dimension b and the sum of their other dimensions is $2a+b$; hence their sum is the rectangle $(2a+b)b$; *i.e.* the rectangle of the sum and difference of the tracts $a+b$ and a. Moreover, this area is plainly what is left on taking away the square on a from the square on $(a+b)$; hence, etc. Q. E. D.

Scholium. Calling the tracts u and v, we may express these theorems in symbols thus: $(u+v)^2 = u^2 + v^2 + 2uv$; $(u-v)^2 = u^2 + v^2 - 2uv$; $(u+v)(u-v) = u^2 - v^2$.

But let the student carefully beware of importing any algebraic meaning into these symbols at this stage of the discussion; u^2, for example, does *not* mean the *product* of u multiplied by u, but merely the *square* whose side is u.

183. Theorem XCV. — *The square on the hypotenuse of a right △ equals the sum of the squares on its other sides* (Fig. 125).

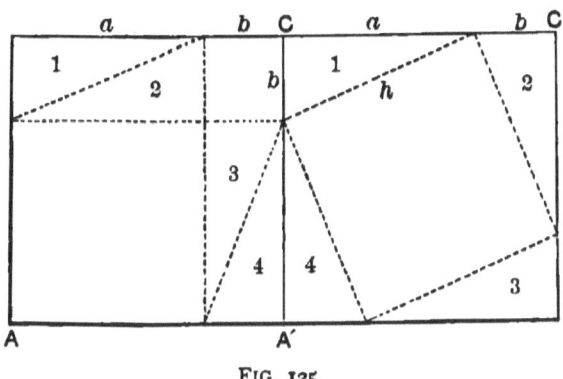

FIG. 125.

Proof. Let AC and $A'C'$ be two congruent squares on the tract $a+b$; take away from each the four congruent right △, 1, 2, 3, 4; there is left of the one the square on the hypotenuse of the right △, and of the other the sum of the squares on the legs, a and b. Q.E.D.

Scholium. This most famous theorem was discovered by Pythagoras (circa B.C. 550), it is said, while pursuing certain arithmetical researches. He was, in fact, seeking out pairs of numbers, the sum of whose squares was itself the square of a number, when he made the astonishing observation that all such pairs, when measured off in terms of a unit length, formed the two legs of a right △ of which the third number, similarly laid off, formed the hypotenuse. Proofs of the proposition abound. In the classic one of Euclid, it is shown that the square on AC = rectangle AD (Fig. 126), the media of comparison being the congruent △ BAE and FAC. Similarly the square on BC = rectangle BD. The

160   GEOMETRY.   [TH. XCV.

demonstration in the text seems to be the most simple and direct that is possible, while its presuppositions are the least possible.

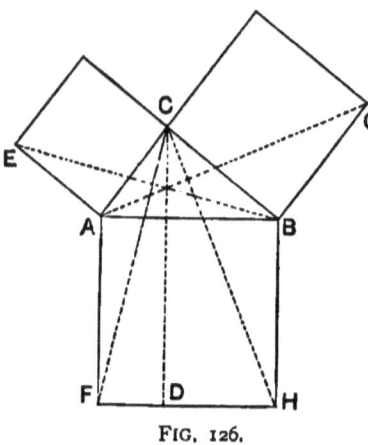

FIG. 126.

184. Naturally we now inquire into the relations among the squares on the sides of oblique △.

Def. The foot of the normal, from a point to a ray, is called the (orthogonal) **projection** of the point on the ray; and the tract between the projections of the ends of a tract is called the **projection** of the tract itself (Fig. 127).

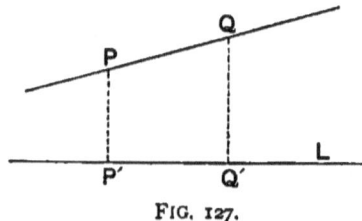

FIG. 127.

Thus P' and Q' are the projections of P and Q on L, and $P'Q'$ is the projection of PQ on L.

SQUARES.

185. Theorem XCVI. — *The square on any side of a triangle equals the sum of the squares on the other sides,* increased *or* decreased *by twice the rectangle of either and the projection of the other upon it, according as the first side lies opposite an* obtuse *or an* acute *angle.*

Data: ABC the $\triangle$, BD the projection of BC on AB (Fig. 128).

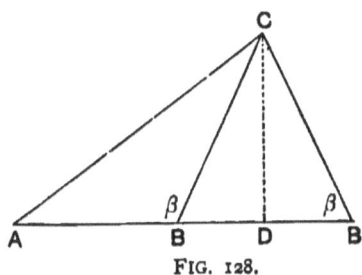

FIG. 128.

Proof. $AC^2 = AD^2 + CD^2$,

and $CD^2 = BC^2 - BD^2$ (why?);

hence $AC^2 = AD^2 - BD^2 + BC^2$.

But $AD^2 - BD^2 = AB(AB + 2BD)$ (why?);

hence $AC^2 = AB^2 + BC^2 + 2AB \cdot BD$.

Thus far β has been considered obtuse; if it be acute, then $AD = AB + BD$; but BD is to be reckoned leftward, oppositely to BD in the former case; that is, BD is reversed in sense, a fact which we may express in symbols by writing

$$AD = AB - DB.$$

Hence results, as before,

$$AC^2 = AB^2 + BC^2 - 2AB \cdot DB. \qquad \text{Q. E. D.}$$

Scholium. We observe that when β is a right angle, then, and then only, D falls on B, the projection on BD vanishes, and there results the Pythagorean Theorem,

$$AC^2 = AB^2 + BC^2.$$

As β changes from obtuse to acute, the point B passes from the left to the right of D, and the tract BD changes its *sense*,— from being reckoned *right*ward it comes to be reckoned *left*ward. It is this change of sense in BD that changes the addition into the subtraction of the rectangle. It is often absolutely necessary to take account of the *sense* of a magnitude, the way it is reckoned, in order to perceive the generality, the internal coherence and continuity, of our results.

Corollary. When the square on one side of a $\triangle$ equals the sum of the squares on the others, the $\triangle$ is right-angled opposite that side. Converse of the Pythagorean Theorem.

186. Theorem XCVII. *The sum of the squares on two sides of a $\triangle$ equals twice the sum of the squares on half the third side and its medial.*

Data: ABC the $\triangle$, CM medial, and CN normal to AB (Fig. 129).

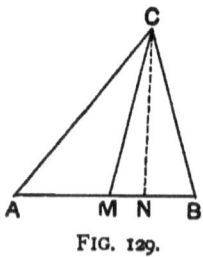

FIG. 129.

Proof. $\overline{AC}^2 = \overline{AM}^2 + \overline{CM}^2 + 2 AM \cdot MN,$

$\overline{BC}^2 = \overline{BM}^2 + \overline{CM}^2 - 2 BM \cdot MN.$

Adding and remembering that $AM = BM$, we get
$$\overline{AC}^2 + \overline{BC}^2 = 2(\overline{AM}^2 + \overline{CM}^2). \qquad \text{Q. F. D.}$$

Corollary 1. If a, b, c be the sides of the $\triangle$, and m the medial of c, then
$$4m^2 = 2a^2 + 2b^2 - c^2.$$

Corollary 2. If the $\triangle$ be regular, then m is the altitude, and if s be a side,
$$4m^2 = 3s^2.$$

*187. **Theorem XCVIII.** — *The sum of the squares on the sides of a quadrangle equals the sum of the squares on the diagonals and four times the square on the join of the mid-points of the diagonals.*

Data: $ABCD$ the 4-side, EF the join of the mid-point of the diagonals (Fig. 130).

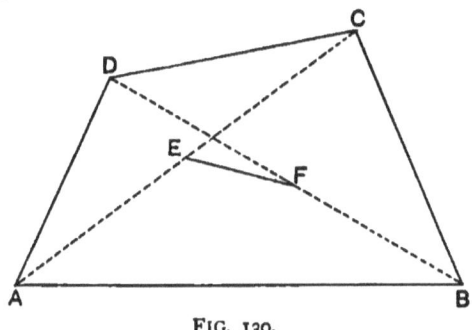

FIG. 130.

Proof. $\overline{AB}^2 + \overline{AD}^2 = 2\overline{AF}^2 + 2\overline{BF}^2$,
$\overline{BC}^2 + \overline{CD}^2 = 2\overline{BF}^2 + 2\overline{CF}^2$ (why?);
whence $\overline{AB}^2 + \overline{BC}^2 + \overline{CD}^2 + \overline{DA}^2$
$= 4\overline{BF}^2 + 2\overline{AF}^2 + 2\overline{CF}^2$
$= 4\overline{BF}^2 + 4\overline{AE}^2 + 4\overline{EF}^2$ (why?)
$= \overline{BD}^2 + \overline{AC}^2 + 4\overline{EF}^2$ (why?). $\qquad$ Q. E. D.

164   GEOMETRY.   [TH. XCIX.

Corollary. The sum of the squares on the sides of a ▱ equals the sum of the squares on the diagonals. (Why?)

188. Theorem XCIX. — *The difference between the squares on a side of a symmetric △ and on the join of the vertex to any point of the base equals the rectangle of the segments into which that point divides the base.*

Data : ABC the symmetric △, P any point of the base (Fig. 131).

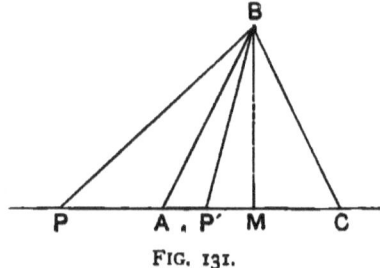

FIG. 131.

Proof. Let M be the mid-point of the base, then

$$\overline{BP}^2 = \overline{BA}^2 + \overline{AP}^2 + 2\,AP \cdot AM \text{ (why?)}.$$

$$\therefore \overline{BP}^2 - BA^2 = AP\{AP + 2\,AM\} = AP \cdot PC. \quad \text{Q. E. D.}$$

Now let P move towards A; as it reaches A, the tract AP vanishes, and so do both sides of the equation. As P moves on to P' towards C, the tract AP *changes sense*, it is no longer reckoned leftward, but rightward; at the same time the left-hand side of the equation *changes sign*, BP becoming less than BA; but the equation still holds, for the sign of AP must also change with the change of sense. We are yet at liberty to choose the difference of squares as either $\overline{BP}^2 - \overline{BA}^2$ or $\overline{BA}^2 - \overline{BP}^2$. The first is perhaps preferable, and we see that when P is without the tract AC, then PA and PC have the same sense and sign, being reckoned the same way, and the difference is positive; but when P is within AC, then PA and PB have opposite sense and sign, being reckoned oppositely; hence we may say their rectangle is negative, and accordingly BP is less than BA. It is extremely important to note that an area is a *sign-magnitude*, positive or negative; it has sense.

189. When P is at A, the difference $\overline{BP}^2 - \overline{BA}^2$ is 0; as P moves towards C, the tract BA remains unchanged, but BP shortens until P reaches M; thence BP lengthens until it again becomes equal to BA or BC, as P falls on C. Hence the difference $\overline{BA}^2 - \overline{BP}^2$, or its equal $PA \cdot PC$ increases while P moves from A to M and decreases as P moves from M to C; hence it is *greatest* when P is at M. A value of a variable magnitude that is thus the greatest within a series of successive values, or that is greater than the values just before it and just after it, is called a **maximum**; while a value that is less than the values next before and next after it, is called a **minimum**. Hence the rectangle $AP \cdot PC$ is a maximum for P at M; or the rectangle on the two parts into which a given tract may be divided is a maximum when the parts are equal, or *of all rectangles with a given perimeter, the square is the maximum*. Once more,

$$\overline{AC}^2 = \overline{AP+PC}^2 = \overline{AP}^2 + \overline{PC}^2 + 2\,AP\cdot PC.$$

Now AC is constant while P moves from A to C, and $AP \cdot PC$ is greatest when P is at M; hence $\overline{AP}^2 + \overline{PC}^2$ is least when P is at M; that is, *the sum of the squares on the two parts of a given tract is a minimum when the parts are equal.*

190. **Theorem C.** — *The rectangle of the distances on a secant from a fixed point to a fixed circle is constant for all secants.*

Data: P the fixed point, S the fixed circle (Fig. 132).

Proof. Through P draw any two secants cutting the circle at A and B and at C and D. Then in the symmetric $\triangle AOB$, $\overline{OP}^2 - \overline{OA}^2 = PA \cdot PB$ and in $\triangle COD$, $\overline{OP}^2 - \overline{OC}^2 = PC \cdot PD$ (why?).

Hence $PA \cdot PB = PC \cdot PD$ (why?), no matter how the secant be drawn through P. Q. E. D.

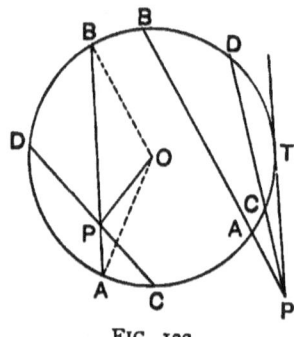

FIG. 132.

Def. This constant, namely, the area of the rectangle of the distances from the fixed point to the fixed circle along any secant, is called the **power** of the point as to the circle.

Corollary 1. For a point within the circle the power is the square on half the shortest chord through the point, or on half the chord through the point normal to the radius through the point (why?); for a point without the circle the power is the square on the tangent-length from the circle to the point (why?); for a point on the circle the power is zero (why?).

Corollary 2. The power of a point without the circle is positive; of a point within, it is negative (why?).

Corollary 3. If PT^2 = power of P as to S, and T be on S, then PT is tangent to S (why?).

Corollary 4. If m be the minimum distance from the point P to the circle S of diameter d, then the power of the point is $\mathbf{m}(\mathbf{d} \pm \mathbf{m})$ according as P is without or within S. This notion of the power of a point as to a circle is so exceedingly important that it may be well to exemplify its use, in passing, though not necessary for our present purposes.

191. Theorem CI. — *All points having equal powers as to two circles lie on a ray.*

Data: S and S' the two circles (Fig. 133).

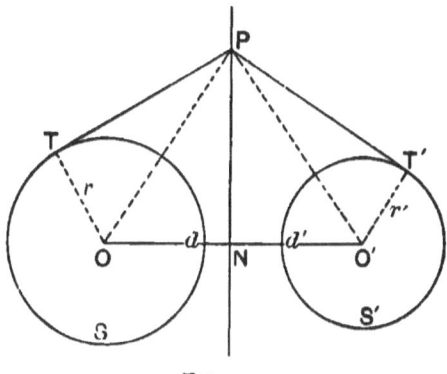

FIG. 133.

Proof. Draw a normal to the centre-tract OO' (at N), cutting it into two parts d and d', and from any point P on this normal draw tangents PT, PT'. Then r and r' being the radii,

$\overline{PN}^2 + d^2 = \overline{PT}^2 + r^2$, and $\overline{PN}^2 + d'^2 = \overline{PT'}^2 + r'^2$ (why?).

Hence $d^2 - d'^2 = \overline{PT}^2 - \overline{PT'}^2 + r^2 - r'^2$.

Hence $\overline{PT}^2 = \overline{PT'}^2$ *when and only when* $d^2 - d'^2 = r^2 - r'^2$.

If then we find N on OO', dividing it so that $d^2 - d'^2 = r^2 - r'^2$, and this can always be done, then the powers of all points on the normal through N will be equal, and the powers of all points not on this normal will be unequal. Q. E. D.

192. *Def.* This most important ray was discovered in 1813 by Gaultier and named by him *radical axis* of the two circles; a better name would seem to be **power-axis**. This discovery marked and in a measure determined the renascence of Geometry.

Corollary 1. The common secant of two circles is their power-axis.

Corollary 2. The common tangent of two circles is their power-axis.

Corollary 3. The power-axes of three circles taken in pairs concur.

Def. The point of concurrence is named *radical centre* or **power-centre** of the three circles.

Corollary 4. A circle about the power-centre with the common *tangent-length as radius intersects the three circles orthogonally.*

The importance of the following discussions can scarcely be overestimated. They are meant to ground firmly and in strict geometric fashion the doctrine of

PROPORTION.

193. If P be any point not on a circle S, PT a tangent, and PAB a secant of S, then we have seen that $PA \cdot PB = \overline{PT}^2$ for all directions of PAB. If a circle I be drawn about P with radius PT, it will cut S orthogonally (why?); then A and B are called **inverse points** as to P, which is called the **centre** of inversion, while I is called the **circle** of inversion.

194. Theorem CII (converse of CI). — *If the rectangle of distances from a point to two points on a ray equal* (in sense as well as sign) *the rectangle of the distances from the point to two points on another ray, both rays going through the point, then the two pairs of points are encyclic.*

Data: P the point, A and B, C and D, the two pairs of points, and $PA \cdot PB = PC \cdot PD$ (Fig. 134).

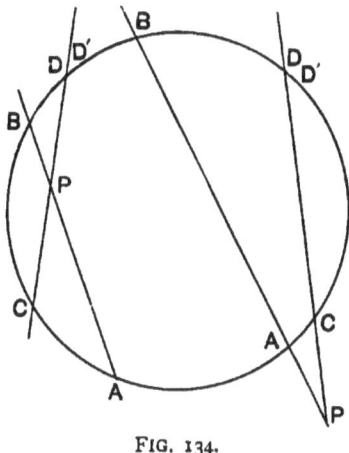

FIG. 134.

Proof. The circle through A, B, and C will meet PD somewhere, as at D'. Then $PA \cdot PB = PC \cdot PD'$ (why?); hence $PD = PD'$ (why?), or D' is D.

Query. Where and why is it necessary to regard the sense of the rectangles in this proof?

195. Now let us consider this encyclic quadrangle. We know that the opposite inner ∠s, as A and D, are supplemental (Fig. 135). Think of the plane as a doubly laid film with AC drawn in the lower and BD drawn in the upper layer, and imagine PBD taken up, turned over, and replaced so that B will fall on B' and D on D'. Then will $B'D'$ be ∥ to AC. For the inner angle at D' equals inner angle at D; hence the inner ∡s at A and D' are supplemental (why?); hence $B'D'$ and AC are ∥.

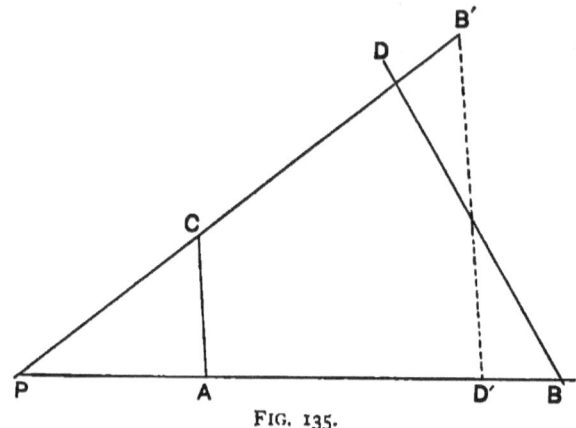

FIG. 135.

This operation of turning over BD into the position $B'D'$ we may call *inverting BD*.

Hence

Theorem CIII. — *If one side of an encyclic quadrangle be inverted, the resulting figure is a trapezoid.*

Conversely,

196. Theorem CIV. — *If one parallel side of a trapezoid be inverted, the figure resulting is an encyclic quadrangle.* The ready proof is left for the student.

We note that the proofs hold as well for the crossed quadrangle and trapezoid as for the convex or normal.

197. Now let PAC and $PB'D'$ be any two △ with common vertical angle P and ∥ bases AC and $B'D'$; then

$$PA \cdot PB' = PC \cdot PD'$$ (Fig. 135).

Proof. For on inverting $B'D'$ into BD the quadrangle $ABCD$ is encyclic. Hence $PA \cdot PB = PC \cdot PD$, and $PB = PB'$, $PD = PD'$. Hence Q. E. D.

198. *Conversely,* Let PAC and $PB'D'$ be any two △ with common vertical ⚹ and let

$$PA \cdot PB' = PC \cdot PD'.$$

Then AC and B'D' are ∥.

Proof. Draw through A a ∥ AC' to $B'D'$, cutting PB' at C'; then

$$PA \cdot PB' = PC' \cdot PD'.$$

Hence $PC = PC'$ (why?). Q. E. D.

199. These relations are very simple and easy of comprehension, but their statement in words is very awkward and cumbrous. To relieve the difficulty of verbal expression we introduce a new arbitrary definition and a new arbitrary symbolism.

Def. If the rectangle of two tracts equals the rectangle of two other tracts, the four tracts are said to be **in proportion**, or to **form a proportion**, or to be **proportional**.

Symbolism. If u and v be the one pair, x and y the other pair of tracts, then we write $u : x :: y : v$ and read *u is to x as y is to v.*

200. In order to speak readily about this proportion we define further:

Definition 1. The tracts are called **terms** of the proportion.

Definition 2. The first and last are called **extremes**; the second and third are called **means**.

Definition 3. The first and second are called the **first couplet**; the third and fourth, the **second couplet**.

Definition 4. The first and third terms are called **antecedents**; the second and fourth are called **consequents**.

Definition 5. When the two means are equal, each is called the **mean proportional** or **geometric mean** of the extremes.

Definition 6. The fourth term is called the **fourth proportional** to the other three taken in order; or, if the means be equal, it is called a **third proportional** to the other two taken in order.

Definition 7. When the means are exchanged, or the extremes are exchanged, the proportion is said to be **alternated**.

Definition 8. When the terms of each couplet are exchanged, the proportion is said to be **inverted**.

Definition 9. When in place of the first or second term of each couplet is put the **sum** (or **difference**) of the terms of that couplet, the proportion is said to be **compounded** (or **divided**).

201. Since by a proportion we mean nothing more and nothing less than that *the rectangle of the means equals the rectangle of the extremes*, it is plain that the same proportion may be written in several different ways, thus:

$$u:x::v:y,\ u:v::x:y,\ y:x::v:u,\ y:v::x:u,$$
$$x:u::y:v,\ x:y::u:v,\ v:u::y:x,\ v:y::u:x,$$

all mean precisely the same; namely, rectangle of u and y = rectangle of v and x.

202. All of these forms, and no others, may be derived from any one of them by alternation and inversion; hence

Theorem CV. — *When four tracts are in proportion they are in proportion by alternation and by inversion* (*alternando et invertendo*).

The simplification of expression will now soon become evident. We must still further premise, however,

Definition 10. When the angles of one △ are respectively equal to those of another, the △ are said to be **mutually equiangular,** and the sides opposite equal angles are said to **correspond,** as do also the equal angles themselves.

203. **Theorem CVI.** — *Corresponding sides in two mutually equiangular △ are proportional in pairs.*
Data: ABC, $A'B'C'$ the △, $A = A'$, $B = B'$, $C = C'$.
Proof. Fit $\angle A$ on $\angle A'$; then BC is ∥ to $B'C'$ (why?).
Hence $AB \cdot A'C' = A'B' \cdot AC$ (why?),
or $AB : A'B' :: AC : A'C'$.

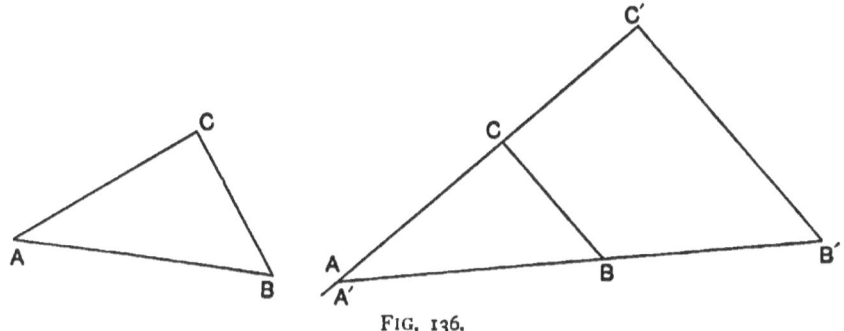

FIG. 136.

Similarly, by putting B on B', C on C',
$BC : B'C' :: BA : B'A'$,
$CA : C'A' :: CB : C'B'$.   Q. E. D.

204. These three proportions may be conveniently written as a **continued proportion,** thus:

$AB : BC : CA :: A'B' : B'C' : C'A'$, read

AB is to BC is to CA as $A'B'$ is to $B'C'$ is to $C'A'$;
or perhaps still better thus:

$AB : A'B' :: BC : B'C' :: CA : C'A'$, read

AB is to $A'B'$ as BC is to $B'C'$ as CA is to $C'A'$;
they are exactly equivalent to the three equations

$$AB \cdot B'C' = A'B' \cdot BC, \quad BC \cdot C'A' = B'C' \cdot CA,$$
$$CA \cdot A'B' = C'A' \cdot AB.$$

205. Theorem CVII. — Conversely, *Two △ with sides proportional are mutually equiangular.*

Data: ABC, $A'B'C'$ the two △, and
$AB : A'B' :: BC : B'C' :: CA : C'A'$ (Fig. 137).

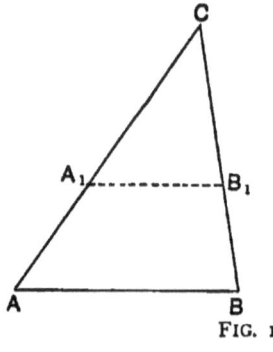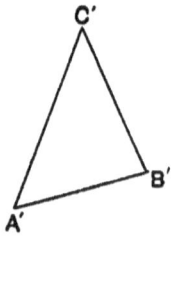

FIG. 137.

Proof. From C lay off CA equal to $C'A'$ and draw $A_1B_1 \parallel$ to AB. Then

$$CA_1 \cdot CB = CA \cdot CB_1 \text{ (why?)}.$$
But $\quad C'A' \cdot CB = CA \cdot C'B'$ (why?).
Hence $\quad CA \cdot CB_1 = CA \cdot C'B'$ (why?).
Hence $\quad CB_1 = C'B'$.
Similarly, $\quad A_1B_1 = A'B'$.

Hence $A'B'C'$ and A_1B_1C are congruent (why?).
Hence $A'B'C'$ and ABC are mutually equiangular (why?).
Q. E. D.

Thus it appears that mutual equiangularity and proportionality of sides *coexist* and *imply each other*. Two such △ mutually equal in their angles and proportional in their sides are called **similar**.

206. Theorem CVIII. — *Two △ having an ∡ of one equal to an ∡ of the other and the including sides proportional are similar.*

Data: ABC, $A'B'C'$ the two △,

$\angle C = \angle C'$, and $CA : C'A' :: CB : C'B'$ (Fig. 138).

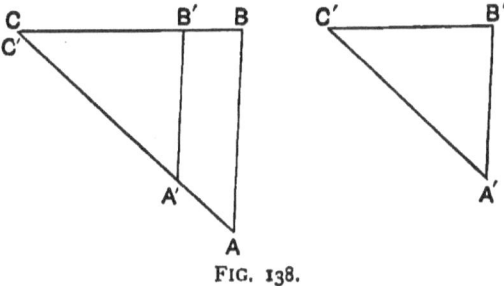

FIG. 138.

Proof. Fit C' on C; then by the proportion $A'B'$ is ∥ to AB. Hence, etc. Q. E. D.

207. Theorem CIX. — *Two △ having two pairs of sides proportional, and a pair of angles opposite the larger sides in each equal, are similar.*

Data: ABC, $A'B'C'$ the two △,

$AB > BC$, $A'B' > B'C'$.

$AB : A'B' :: BC : B'C'$, and $\angle C = \angle C'$ (Fig. 139).

Proof. Fit ∡ C on C'; then since

$A'B' > B'C'$ and $AB > BC$, both AB and $A'B'$ must be drawn making the inner ∡s at A and A' acute. From the proportion, AB and $A'B'$ are now ∥ (prove it); hence the △ are mutually equiangular and hence similar. Q. E. D.

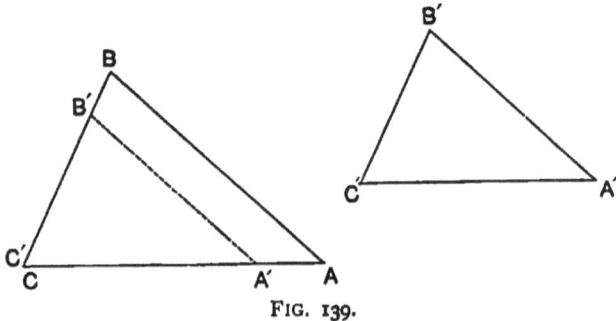

FIG. 139.

N.B. If AB were $< BC$ and $A'B' < B'C'$, then AB and $A'B'$ might make the ∡s at A and A' supplemental instead of equal, and hence AB and $A'B'$ anti-∥ instead of ∥, in which case the △ would not be similar. Draw a figure illustrating this case.

Compare the conditions of similarity with the conditions of congruence between two △.

208. Theorem CX. — *If two proportions agree in the first three terms of each, they agree in the fourth also.*

Data: $u : x :: v : y$, and $u : x :: v : y'$.

Proof. $uy = vx$, and $uy' = vx$; hence $uy = uy'$.

Hence $y = y'$ (why?). Q. E. D.

209. Theorem CXI. — *If two proportions agree in one couplet, the other couplets form a proportion.*

PROPORTION.

Data: $u:x::v:y$, $u:x::w:z$ (Fig. 140).

Proof. Suppose $w < v$ and on two half-rays through P lay off PU, PX, PV, and $PY = u, x, v, y$. Open the angle at P until $UV = w$. This is possible, since $w < v$. Draw XY and call it z'. Then since $u:x::v:y$, w and z' are $\|$; hence $u:x::w:z'$. Hence $z' = z$ (why?), and hence $v:y::w:z'$ or z (why?). Q. E. D.

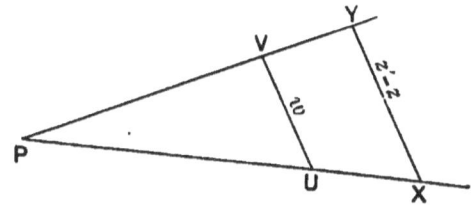

FIG. 140.

Corollary. We may write the three proportions as a continued proportion, thus:

$$u:x::v:y::w:z.$$

210. Theorem CXII. — *If four tracts are in proportion, they are in proportion by composition, and by division, and by composition and division (Componendo, dividendo, componendo et dividendo).*

Datum: $u:x::v:y$.

Proof. On any pair of half-rays through any point P lay off $PU = u$, $PV = v$; from U lay off on P a tract $UX = x$, and on a $\|$ to PV a tract $UY = y$. Then the $\triangle$ PUV and

UXY are similar (why?); hence UV and XY are ∥ (why?), and VR is y (why?). Also △ PUV and PXR are similar (why?). Hence $u : u + x :: v : v + y$ (*Componendo*).

Again, from P lay off as before $PU = u$, $PV = v$, $PX = x$, $PY = y$; then XY and UV are ∥ (why?) (Fig. 141).

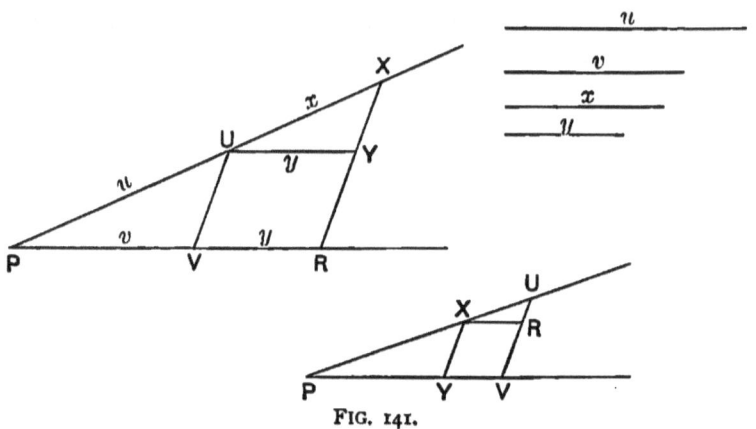

FIG. 141.

Draw XR ∥ to PV; then △ PUV and XUR are similar (why?). And $XR = v - y$ (why?), $XU = u - x$.

Hence $u : u - x :: v : v - y$ (*Dividendo*).

Hence $u + x : u - x :: v + y : v - y$ (why?) (*Componendo et dividendo*).

*211. **Theorem CXIII.** — *When four tracts are in proportion, equimultiples of either or both couplets are in proportion.*

Datum: $\qquad u : x :: v : y.$

Proof. By *Def.* rect. uy = rect. vx. Hence m (rect. uy) = m (rect. vx).

But m (rect. uy) = rect. $mu \cdot y$ and m (rect. vx) = rect. $v \cdot mx$; hence $mu \cdot y = v \cdot mx$, or $mu : mx :: v : y$.

Similarly, $\qquad nv : ny :: v : y.$

Hence $\qquad mu : mx :: nv : ny.$ Q. E. D.

Corollary. The multipliers m and n may be disposed any way in the proportion provided only that each appears in a mean and in an extreme.

212. *Def.* Two tracts are said to be **divided similarly** when all the parts of the one and all the parts of the other taken in the same order form a continued proportion; thus, if a, b, c, d be the parts of one and a', b', c', d' the parts of the other, and $a : b : c : d : : a' : b' : c' : d'$, then the divison is *similar*. Two parts forming a couplet, as a and a', are said to *correspond*.

213. Theorem CXIV. — *Two transversals of a system of parallels are divided similarly by the parallels.*

Data: AA', BB', etc., the ∥s, PT and PT' the transversals (Fig. 142).

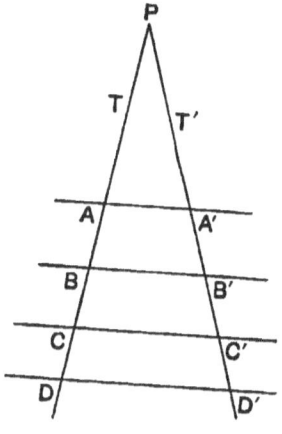

FIG. 142.

Proof. From similar △,
$$PA : PA' : : PB : PB' : : PC : PC' : : PD : PD'.$$
Hence, *alternando* and *dividendo*,
$$PA : PA' : : AB : A'B' : : BC : B'C' : : CD : C'D'. \quad \text{Q. E. D.}$$

Corollary 1. The intercepts of the ∥s are proportional to their distances from the vertex P.

Corollary 2. A system of ∥s divides the rays of a pencil similarly (Fig. 143).

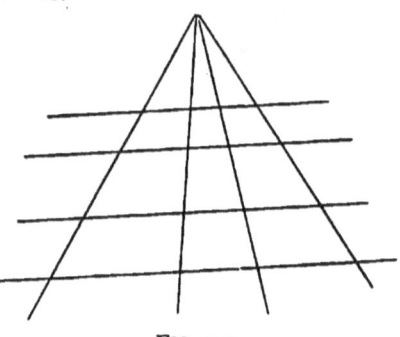

FIG. 143.

Corollary 3. The intercepts of the rays on any two ∥s are proportional.

*214. **Theorem CXV** (Ptolemy's). — *In an encyclic quadrangle the rectangle of the diagonals equals the sum of the rectangles of the opposite sides.*

Datum: $ABCD$ an encyclic quadrangle.

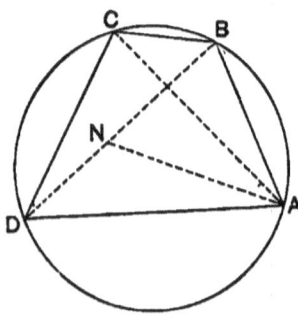

FIG. 144.

Proof. Draw the diagonals and also AN, making $\angle AND = \angle ABC$. Then $\triangle ADN$ and ACB are similar (why?); hence
$$BC \cdot AD = AC \cdot DN \text{ (Fig. 144)}.$$
Also BNA and CDA are similar (why?); hence
$$AB \cdot CD = AC \cdot BN.$$
On addition there results
$$AB \cdot CD + BC \cdot DA = AC \cdot BD. \qquad \text{Q. E. D.}$$

*215. **Theorem CXVI.** — *The rectangle of two sides of a $\triangle$ equals the rectangle of the altitude to the third side and the circum-diameter.*

Data: ABC the $\triangle$, S the circumcircle (Fig. 145).

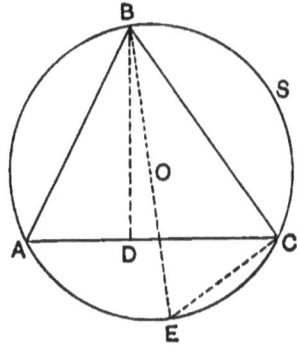

FIG. 145.

Proof. $\triangle ABD$ and EBC are similar (why?).

Hence $AB : EB :: BD : BC$,

or $AB \cdot BC = EB \cdot BD.$ \qquad Q. E. D.

216. *Def.* When a tract is divided into two parts proportional to two other tracts, it is said to be divided **in the ratio**

of those tracts, or the **ratio** of the parts is said to **equal** the **ratio** of the tracts.

N.B. This is a definition of **equality of ratios,** but not of *ratio* itself; this latter we now neither need nor attempt to define, but we write it thus, $l : m$, and read *ratio of l to m*.

Def. The division may be *inner*, when the dividing point P falls within the tract, or *outer*, when it falls without the tract.

217. Theorem CXVII.—*A tract may be divided innerly and outerly in any given ratio, but in each case at only one point.*

Data: AB the tract to be divided, l and m the other tracts.

Proof. 1. From A draw any half-ray; lay off on it $AL = l$ and $LM = m$; join BM, and draw PL ∥ to it (Fig. 146).

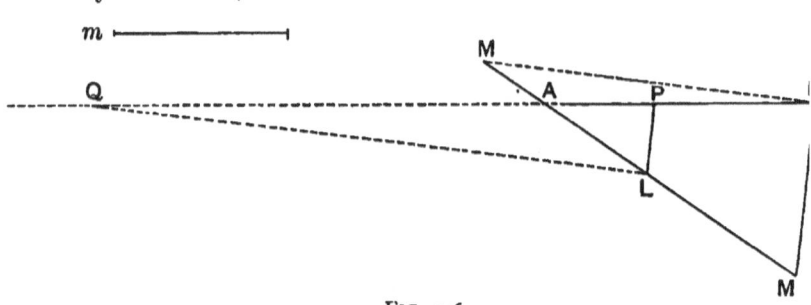

FIG. 146.

Then $AP : PB :: l : m$ (why?); hence P divides AB *innerly* in the given ratio.

Again supposing $m < l$, lay off LM backwards towards A; join BM and draw LR ∥ to it.

Then $AQ : QB :: l : m$ (why?); hence Q divides AB *outerly* in the given ratio.

Proof. 2. If P' be any point of division of AB in the ratio $l:m$, then LP' is ∥ to MB (why?); but there is only one ∥ to AB through L; hence P' falls on P, *i.e.* there is only one point of inner division in the ratio $l:m$. Similarly, show that there is only one point Q of outer division in the ratio $l:m$. Q.E.D.

218. N.B. 1. In case of inner division the parts AP, PB are reckoned the same way, both rightward; in case of outer division the parts AQ, QB are reckoned oppositely, one rightward, the other leftward.

2. In speaking of the ratio of the tracts l and m the order of mention is essential; the ratio of l and m (whatever it may be) is not the same as the ratio of m and l. So, too, the order of mention of the ends of the tract AB is essential: we mean that the first part is to be reckoned *from* A and the second part *to* B; to divide AB in a given ratio is not the same as to divide BA in that ratio.

Def. When a tract AB is divided innerly and outerly at P and Q in the same ratio, it is said to be divided **harmonically**, A and B are said to be **harmonically conjugate** with P and Q, A and B, P and Q are said to form two **harmonic pairs**, and the four points A, P, B, Q, taken in order, are said to be four **harmonic points** or to form an **harmonic range**.

219. **Theorem CXVIII.** — *If P and Q divide AB harmonically, then A and B divide PQ harmonically.*

Data: AB a tract, P and Q the points of inner and outer division in any ratio, as $l:m$.

Proof. $AP:PB :: l:m$, and $AQ:QB :: l:m$;

hence $\quad AP:BP :: AQ:QB$ (why?),

or $\quad PA:AQ :: PB:BQ$ (why?). Q.E.D.

220. N.B. 1. To the inner and outer division of AB by P and Q corresponds the outer and inner division of PQ by A and B.

2. The term *harmonic* is borrowed from the theory of musical intervals; four tracts a, b, c, d are said to be harmonically or musically proportional when the first is to the last as the difference of the first two is to the difference of the last two; *i.e.* when $a : d :: a - b : c - d$.

Now let the student prove that if $AP: PB :: AQ: QB$,

then $\quad AP: QB :: AP - PB : AQ - QB$,

and so justify the use of the term *harmonic*.

221. Theorem CXIX. — *The inner and outer mid-rays of an angle of a $\triangle$ divide the opposite side harmonically in the ratio of the adjacent sides.*

Data: CP and CQ, the inner and outer mid-rays of the angle C of the $\triangle ABC$, meeting the side AB at P and Q (Fig. 147).

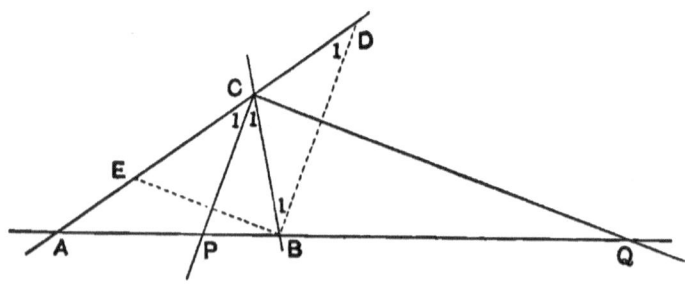

FIG. 147.

Proof. Draw BD ∥ to CP; then $\angle DBC = \angle BCP$ (why?) $= \angle PCA$ (why?) $= \angle CDB$ (why?); hence $CD = CB$.

Also $AP:PB::AC:CD$ (why?) or $AP:PB::AC:CB$.
Similarly, $AQ:QB::AC:CB$.
Hence $AP:PB::AQ:QB::AC:CB$. Q.E.D.

Corollary. Conversely, if two rays divide a side of a △ innerly and outerly in the ratio of the adjacent sides, they are the inner and outer mid-rays of the opposite angle (why?).

222. Theorem CXX. — *If a normal be drawn from the vertex of the right angle in a right △ to the hypotenuse, then*

I. *The △ will be cut into two right ▲ similar to each other and to the original △.*

II. *The normal tract will be a mean proportional between the segments of the hypotenuse.*

III. *Each side of the right angle will be a mean proportional between the whole hypotenuse and the adjacent segment.*

Data: ABC the right △, CN the normal to the hypotenuse (Fig. 148).

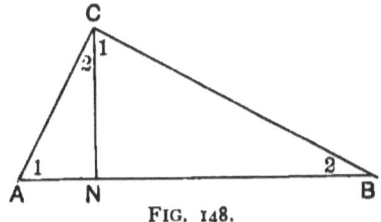
FIG. 148.

Proof. I. The ▲ ABC, ACN, and BCN are plainly mutually equiangular (why?) and hence similar. Q.E.D.

II. Hence $AN:CN::CN:NB$. Q.E.D.

III. Also $AB:AC::AC:AN$,
and $AB:BC::BC:BN$. Q.E.D.

Corollary. Conversely, if a tract CN from the right angle divides the $\triangle$ into similar $\triangle$, or is a mean proportional between the segments of the hypotenuse, or divides the hypotenuse so that either side is a mean proportional between the whole hypotenuse and the adjacent segment, then it is normal to the hypotenuse.

223. Theorem CXXI. — *If four concurrent rays (or rays of a pencil) cut one transversal harmonically, they cut every transversal harmonically.*

Data: Any transversal cut harmonically at A, B, C, D by four rays concurrent in O and $PQRS$ any other transversal (Fig. 149).

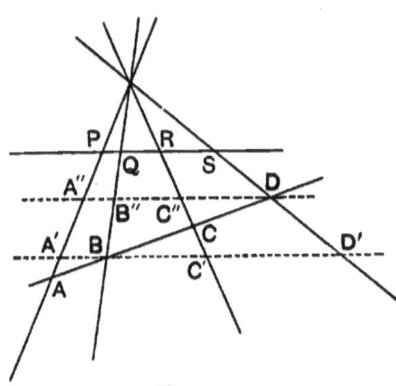

FIG. 149.

Proof. Draw through two conjugate points as B and D any two transversals ‖ to $PQRS$; then

$$AB : BC :: AD : DC,$$

or $\quad AB : AD :: BC : DC$ (why?).

Also $\quad AB : AD :: A'B : A''D$,

and $\quad BC : DC :: BC' : DC''$ (why?);

hence $\quad A'B : A''D :: BC' : DC''$,

or $\quad A'B : BC' :: A''D : DC''$.

But $\quad A''D : DC'' :: A'D' : D'C'$ (why?);

hence $\quad A'B : BC' :: A'D' : D'C'$;

hence $\quad PQ : QR :: PS : SR$ (why?). Q. E. D.

Corollary. The proportion $AB : BC :: AD : DC$ is not affected by any movement of O, while A, B, C, D remain fixed; neither, then, is the proportion $PQ : QR :: PS : SQ$.

224. **Theorem CXXII.** — *A chord of a circle and the tangents at its ends cut the conjugate diameter harmonically* (Fig. 150).

Data: S the circle, TT' the chord, PT, PT' tangents at its ends, PA the conjugate diameter.

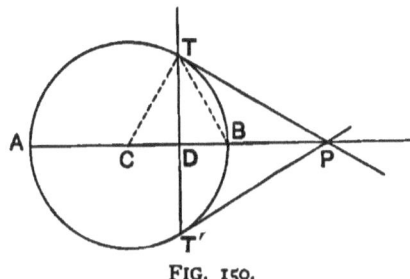

FIG. 150.

Proof. TB bisects $\angle PTD$ innerly (why?);

hence $\quad TA$ bisects it outerly (why?);

hence $\quad A$, D, B, P are four harmonic points (why?).

Q. E. D.

SIMILAR FIGURES.

225. We have already found that mutually equiangular △ have their corresponding sides proportional, and conversely, and we have named such △ similar. A more general notion of similarity may be obtained thus:

Let two points, P and P', move at will in the plane, but under these restrictions:

1. The ray PP' shall pass always through a fixed point O.
2. The proportion shall always hold $OP: OP' :: t: t'$, where t and t' are any two fixed tracts; then the paths of P and P' are called **similar figures similarly placed** (or **homothetic**).

The point O is called the **centre of similitude**; *outer*, if O divides PP' *outerly; inner*, if *innerly*.

226. If an eye were placed at the outer centre, it would manifestly see the one figure through the other, point for point; hence the two figures are said to be in *direct* **perspective**; if O be the inner centre, they may be said to be in *indirect* **perspective,** or in contra-perspective.

If on any ray through O there be taken two points, C and C', such that $OC: OC' :: t: t'$, then C and C' are said to *correspond* to each other with respect to the centre of similitude O in the *ratio* of similitude $t: t'$; any tract between two points in the one figure is said to *correspond* to the tract between the corresponding points in the other figure.

227. Theorem CXXIII. — *Correspondent tracts in two perspective figures are parallel and in the ratio of similitude to each other.*

Data: O the centre, A and A', B and B' two pairs of corresponding points (Fig. 151).

TH. CXXV.] SIMILAR FIGURES. 189

Proof. The △ AOB and $A'OB'$ are similar (why?); hence AB and $A'B'$ are ∥, and $AB : A'B' :: t : t'$ (why?). Q. E. D.

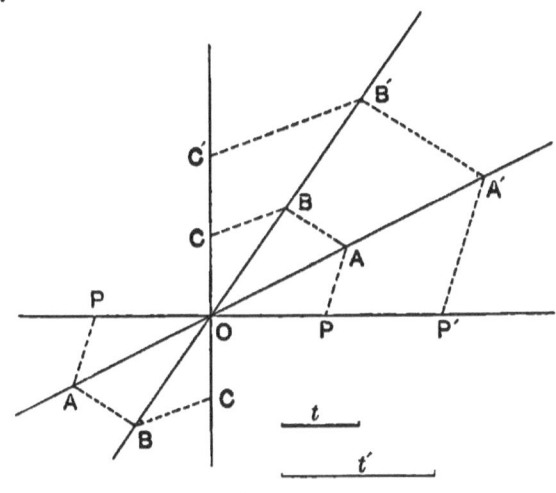

FIG. 151.

228. Theorem CXXIV. — Conversely, *If from two points, A and A', correspondent as to O, there be laid off two ∥ tracts AB, A'B' in the ratio OA : OA', then B and B' correspond.* Let the student give the proof.

229. Theorem CXXV. — *Any two circles are in perspective and contra-perspective.*

Data: S and S' any two circles (Fig. 152).

Proof. Divide the centre tract CC' innerly and outerly in the ratio of the radii $r : r'$ at points I and O. Draw any secant OA, and on it lay off OA' so that

$$OC : OC' :: OA : OA'.$$

The △ OCA and $OC'A'$ are similar (why?). Hence $C'A' = r'$ (why?); hence A' is on S'; hence any point of

S has its correspondent on S' in the same fixed ratio of the radii; hence S and S' are similar, and are plainly in perspective. For I the reasoning is the same, but the tracts being laid off oppositely, the figures are in contra-perspective.

Q. E. D.

Corollary. Common tangents to the two circles go each through a centre of similitude.

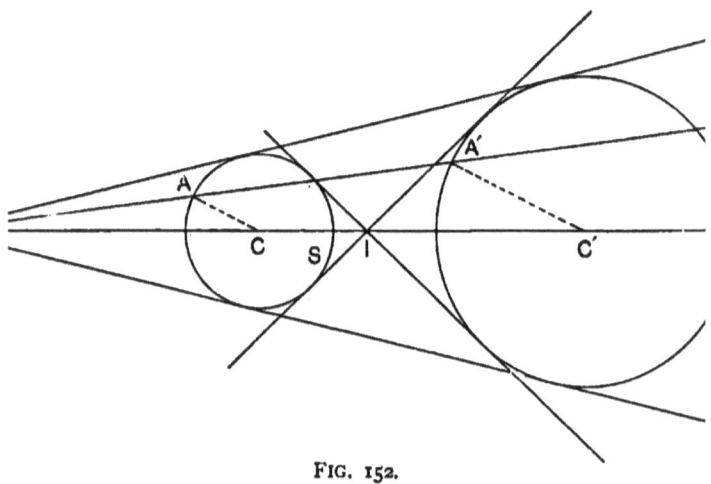

FIG. 152.

230. Theorem CXXVI. — Conversely, *Any figure similar to a circle is itself a circle.*

Data: S a circle, S' similar to it with respect to the centre of similitude O, in the ratio $r : r'$.

Proof. Find the corresponding point C' of the centre C of S, and draw through O any secant meeting S and S' in the corresponding points A and A'. Then triangles COA and $C'OA'$ are similar (why?); hence

$$OC : OC' :: CA : C'A',$$

Tн. CXXVIII.]   SIMILAR FIGURES.   191

and since OC, OC', and CA are constant in length, so, too, is $C'A'$. Hence S' is a circle about C' as centre. The like proof holds for the inner centre I. Q.E.D.

231. **Theorem CXXVII.** — *The angle between two tracts in one figure equals the angle between the corresponding tracts in any similar figure.*

Data: F and F' (Fig. 151), two similar figures similarly placed. AB and BC, two tracts in F. $A'B'$ and $B'C'$, the corresponding tracts in F'.

Proof. Draw OA, OB, OA', OB'; then the theorem follows at once from similar △. But if the figures be not similarly placed, and F'' be one of them congruent with F', suppose F'' brought to coincidence with F'; then what has just been proved for F' holds for F''. Q.E.D.

Corollary. F'' may be brought to coincide with F' by being merely *pushed*,— all rays remaining parallel to themselves in their original position, until one point of F'' falls on the corresponding point F', — and then being merely *turned* until another point of F'' falls on its correspondent in F'. If the figures still do not coincide throughout, but only on the common ray through three points, it will be necessary and sufficient to revolve F' about that common ray through a straight angle, which revolution will change opposition into superposition of the figures. In this revolution all rays in the figure are turned through the same angle; hence

232. **Theorem CXXVIII.** — *In two similar figures all lines are inclined to their correspondents under the same angle.* Perhaps we might name this angle the *anomaly* of the one figure as to the other.

In two similar figures point corresponds to point, angle to

equal angle, tract to tract, in the same ratio; hence it is plain that

233. Theorem CXXIX. — *Any two similar figures may be cut up into pairs of similar figures in the same ratio of similitude and order of arrangement.*

INSTRUMENTS.

234. There are four important instruments used in practice in the construction of similar figures: proportional compasses, sector, diagonal scale, and Pantagraph or Eidograph. Of these the last is the most interesting and illustrates in its working very accurately the definition given above of similar figures in contra-perspective. Every well appointed academy should be furnished with these instruments, which may easily be explained and operated.

CONSTRUCTIONS.

235. The doctrine of proportion is extensively employed, not only in mechanical drawing with the instruments mentioned, but also in the strict logical solution of problems of construction.

236. Problem I. — *To divide a tract (innerly and outerly) in a given ratio, as of $l:m$.* (See p. 182.)

237. Problem II. — *To divide a tract (innerly and outerly) into any number of parts proportional to $l, m, n, p, \cdots$.*

Solution. From the beginning of the tract AB draw any half-ray, as AR; on it lay off in order consecutively the tracts $l, m, n, p, \cdots$. Join the end of the last with the end of AB, and through the ends of the others draw parallels; to this join RB; they divide AB as required (why?).

Let the student solve the problem of outer division.

238. Problem III. — *To construct the geometric mean of two tracts.*

Solution. On the sum of the two tracts, l and m (Fig. 153), as diameter, draw a circle, and through their common point draw a half-chord conjugate to the diameter; it is the mean proportional required (why?).

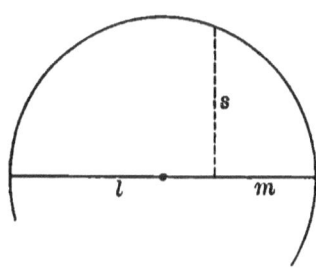

FIG. 153.

239. Problem IV. — *To construct a square equal to a given rectangle.* Proceed as in Problem III.

240. Problem V. — *Knowing one dimension of a rectangle equal to a given rectangle, to find the other; or, given three tracts, to find a fourth proportional to them in order.*

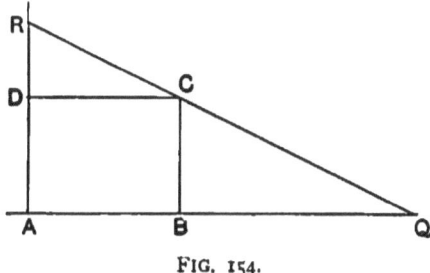

FIG. 154.

Solution. Prolong one side of the given rectangle by the given side of the other, as AB to Q, and draw QC meeting

AD at R; then DR is the other dimension sought (why?) (Fig. 154). Or,

On any two half-rays meeting at A lay off AB and AD equal to the given sides or the second and third of the three tracts, and on either, as AD, lay off AQ equal to the one given dimension or the first tract. Draw BQ and $DR \parallel$ to BQ; then AR is the fourth proportional sought (why?).

These constructions require us either to know the angle at A or else to draw a parallel. But we may proceed thus, *avoiding all use of parallels and angles:* draw a large circle and lay off as a chord of it the difference of the second and third proportionals; from the ends A and B of this chord lay off AP and BP equal to the second and third proportionals; about P describe a circle with the first proportional as radius intersecting the circle at I; draw PI, meeting the circle also at J; then PJ is the fourth proportional sought (why?) (Fig. 155).

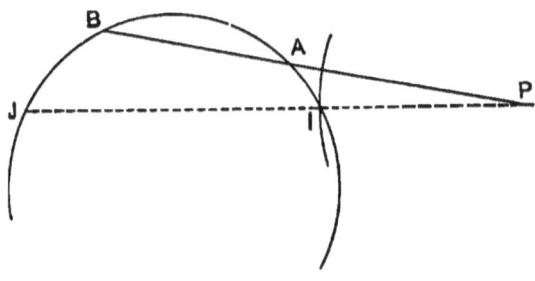

FIG. 155.

N.B. The first circle must be drawn sufficiently large, so that the second circle may meet it.

241. Problem VI. — *To construct a △ equal to a given 4-side.*

Solution. Drawn either diagonal, as AC, of the 4-side $ABCD$, and then from D draw a ‖ to the diagonal, cutting AB at A'. Then $A'CB$ is the △ sought (why?) (Fig. 156).

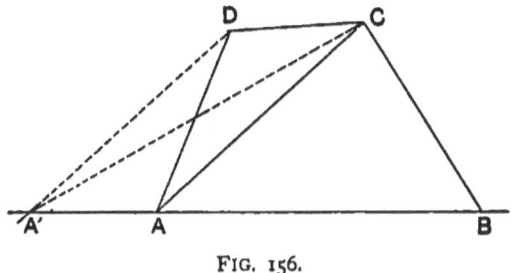

FIG. 156.

242. Problem VII. — *To construct a △ equal to a given n-side.*

Proceed as in Problem VI, and reduce one by one the number of sides drawn to three.

243. Problem VIII. — *To divide a parallelogram into n equal parts by parallels to a side.*

Solution. Divide an adjacent side into n equal parts and draw parallels; the n resulting parallelograms are congruent (why?).

244. Problem IX. — *To divide a △ into n equal parts by tracts drawn from a vertex.*

245. Problem X. — *To divide a trapezoid into n equal parts by tracts between and parallel to the parallels.*

246. Problem XI. — *To divide a △ into n equal parts by tracts drawn from a point on a side.*

Solution. Divide the side containing the point into n equal parts; from the points of division draw parallels to the join of the point with the opposite vertex; draw tracts from the point to the intersections of these parallels with the other sides; they are the dividing lines required (why?).

247. Problem XII. — *From a point within a △ to bisect the △ by tracts drawn to a given vertex and to a side* (Fig. 157).

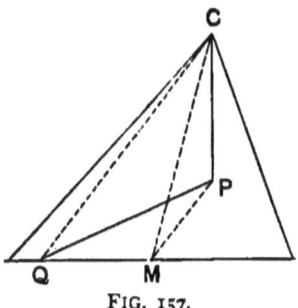

FIG. 157.

Solution. If P be the given point, C the given vertex, and PQ the required tract, then on drawing the medial CM it becomes plain that $\triangle CPQ = \triangle CMQ$; hence CQ is ∥ to PM. Hence the construction: draw the medial CM, then PM, then CQ ∥ to PM, then PQ.

248. Problem XIII. — *From a point within a △ to bisect the △ by two tracts, one of which is drawn to a given point on one side, and the other as may be* (Fig. 158).

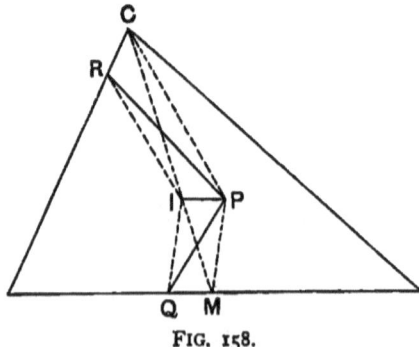

FIG. 158.

CONSTRUCTIONS. 197

Solution. If P be the given point within the $\triangle$, Q the given point on the side, suppose PR to be the required tract. Then on drawing CM and PM and a ∥ to PM through Q cutting CM at I, we have

$$\triangle PQM = \triangle PIM \text{ (why?)}.$$

Also on drawing CP and IR we must have

$$\triangle PIC = \triangle PRC \text{ (why?)}.$$

Hence IR is ∥ to PC (why?).
Hence we construct PR (how?).

249. Problem XIV. — *To bisect an n-side by a tract drawn from a given vertex.*

Solution. Let A be the given vertex; join the adjacent vertices B and L, and through each of the others draw a tract across the n-side ∥ to BL. Bisect these parallels by a train of tracts from A. This broken line will bisect the n-side (why?), and by Problem VII the student may convert it into a single tract from A (how?).

250. Problem XV. — *To construct a square equal to the sum of two given squares.*

Use the Pythagorean Theorem.

251. Problem XVI. — *To construct a square equal to the sum of two given rectangles.*

Combine the methods of Problems IV and XV.

252. Problem XVII. — *To construct a square equal to 2, 3, 4, $\cdots$ n times a given square* (Fig. 159).

Hint. The diagonal of the given square will be the side of the double square (why?); normal to this diagonal, OB, lay off BC equal to the side of the square; draw OC, and again normal to it lay off CD equal to the side of the original square; draw OD, and so on. In this way we may duplicate, triplicate, n-plicate the original square. The broken line $ABCD \cdots$ and the varying hypotenuse wind round O forever.

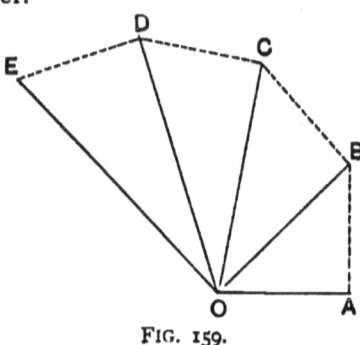

FIG. 159.

253. Problem XVIII. — *To construct a square equal to one half, one third, one fourth, $\cdots$ one nth of a given square.*

Hint. Half the diagonal of the given square is the first side sought; the altitude of a regular triangle whose side is the given side is the second; one half of the given side is the third; $\cdots$ in general, the geometric mean between the side and the nth part of the side of the given square will be the side of the square sought (why?).

254. Problem XIX. — *To construct a square the* nth-*fold or the* nth *part of a given rectangle, parallelogram, or* △.

Combine the methods of the foregoing problems.

255. Theorem CXXX[a]. (Lemma). — *If a rectangle equal a square, and the dimensions of the two be changed proportionally (i.e: so that the new and the old dimensions taken in pairs of correspondents form a continued proportion), then the new rectangle will equal the new square.*

Data: R and R', two rectangles with dimensions a and b, a' and b', s and s', two squares with dimensions t and t'; $R = S$, and $a : a' :: b : b' :: t : t'$ (Fig. 160).

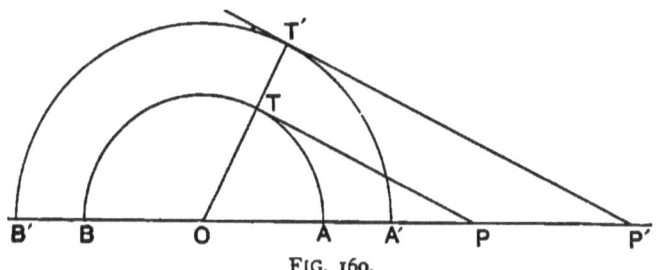
Fig. 160.

Proof. On the same half-ray from the same point P lay off tracts PA and PB equal to a and b; on their difference ($AB = 2\ r$) as diameter describe a circle and draw the tangent PT: it will equal t (why?). With a radius r' such that $a : a' :: r : r'$ describe a concentric circle, prolong OT to meet this circle at T', and at T' draw a tangent meeting the diametral ray at P'.

Then $P'A' = a'$, $P'B' = b'$, $P't' = t'$ (why?),

and $a'b' = t't'$ (why?). Q. E. D.

Corollary. If two rectangles (or parallelograms or △) be equal, and their dimensions be changed proportionally, they will remain equal.

Prove this corollary in detail, and state it along with the foregoing theorem in symbols.

256. Problem XX. — *To construct a rectangle similar to a given rectangle but of double the area.*

Solution. Construct a square equal to the given rectangle; then either dimension of the double rectangle will be a

fourth proportional to the side and diagonal of the square and the corresponding dimension of the given rectangle; that is,
$$s : d :: a : a' \text{ (why?)}.$$

Now, however, all squares are similar (why?); hence, denoting by d' the diagonal of the square on the side a, we have $s : d :: a : d'$; whence $a : d' :: a : a'$, or $a' = d'$.

Similarly, b' is the diagonal of the square on the other dimension b. Or we may find b' by drawing a diagonal of the double rectangle through the end of a' parallel to the diagonal through the end of a.

257. Problem XXI. — *To construct a rectangle similar to a given rectangle but of n-fold the area.*

Solution. If we construct a square, of side s, equal to the given rectangle and also a square, of side s', equal to the required rectangle, then if a and a' be corresponding dimensions in the two rectangles, we have
$$s : a :: s' : a' \text{ (why?)}.$$

Now, however, if we construct, according to Problem XI, two broken lines, one on s as a basis, the other on a as basis, the two will be similar figures (why?); so that if s_n and a_n be corresponding hypotenuses in the two figures, we have
$$s : a :: s_n : a_n.$$
Hence $\qquad s' : a' :: s_n : a_n;$
hence, if $\qquad s_n = s'$, then $a_n = a'$.

Accordingly, we find a' by constructing (Problem XI) the side of a square the n-fold of the square on a. The construction is then completed (how?).

CONSTRUCTIONS. 201

258. **Problem XXII.** — *Let the student extend the same methods to the construction of parallelograms and △ similar to given ones but of n-fold area.*

259. **Problem XXIII.** — *To construct a rectangle (parallelogram or △) similar to a given one but of one half, one third · · · one nth the area.*

260. **Problem XXIV.** — *To construct a figure similar to a given figure but of double, triple, · · · n-fold area.*
Solution. On any tract in the figure construct a square, then construct another square, of double, triple, · · · n-fold area; its side will be the tract in the new figure corresponding to the assumed tract in the original figure (why?). All other points and lines in the required figure may now be found by drawing parallels. Let the student carry out the construction.

261. **Problem XXV.** — *To construct a figure similar to a given figure but of one half, one third, · · · one nth the area.*
The solution is like that of Problem XVIII, *mutatis mutandis*.

262. We have learned to inscribe in a circle a regular 3-side, 6-side, 12-side, · · · $3 \cdot 2^n$-side, also a regular 2^n-side, and it is natural to inquire how to inscribe a regular 5-side, 10-side, · · ·, $5 \cdot 2^n$-side. As it was easiest to begin with the 6-side, so it is easiest to begin with the 10-side; to inscribe the 5-side directly presents difficulties.

263. **Problem XXVI.** — *To inscribe a regular* 10-*side in a circle,* suppose the problem solved and AB the side sought. Then in the symmetric △ AOB the vertical ∠ is half of either basal angle (why?) (Fig. 161).
Hence, if we draw AC bisecting angle A the △ AOB and BAC will be similar (why?).

Hence $OB:AB::AB:BC$, or $OB:OC::OC:BC$.

Hence, in order to find AB or OC, it is necessary to *divide the radius into two parts of which one is the geometric mean of the whole and the other.* This celebrated section is

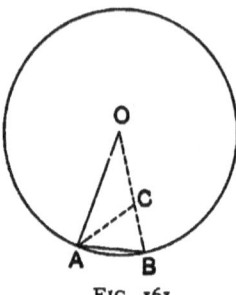

FIG. 161.

called the **median** or **golden section**, and the radius (or any tract so divided) is said to be divided in **extreme and mean ratio**.

The problem of inscribing the 10-side is reduced then to the following:

264. Problem XXVII. — *To divide a tract in extreme and mean ratio.*

Solution. Let a be the tract, and b the greater part; then $\quad a:b::b:a-b$ (Fig. 162),

or $\quad a:a+b::b:a$ (why make this change?).

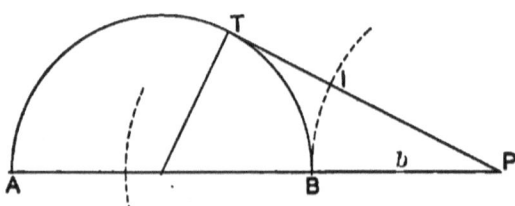

FIG. 162.

CONSTRUCTIONS. 203

Hence we may conceive of a^2 as the power of a point whose distances from the circle (along a diameter) are $a + b$ and b (why?), so that a is the diameter of the circle. Hence, on a as diameter draw a circle; at any point T of the circle draw a tangent and on it take $TP = a$; draw the diameter PBA; then $PB = b$ (why?), and the arc about P with radius b divides PT or a at I in extreme and mean ratio. Q. E. F.

N.B. To the point I corresponds the harmonic conjugate O, the point of *outer* median section, such that

$$PI : IT : : PO : OT.$$

Let the student show that

$$PT \cdot TO = \overline{PO}^2,\ \text{just as}\ PT \cdot TI = \overline{PI}^2.$$

How shall we now construct a regular 5-side, 20-side, $\cdots$ $5 \cdot 2^n$-side?

By combining the constructions for a 3-side and a 5-side we may now construct a regular 15-side. For the difference of the arcs subtended by a side of a regular 3-side and a side of a regular 5-side, is $(\frac{1}{3} - \frac{1}{5})$ of a circle, or $\frac{2}{15}$ of a circle; half of it is $\frac{1}{15}$, or $(\frac{1}{6} - \frac{1}{10})$ of a circle, that is, the arc subtended by one side of a regular 15-side. Hence solve

Problem XXVIII. — *To construct a regular* $15 \cdot 2^n$-*side*.

265. At this point the query seems to arise naturally: if we can find the arc of a side of a 15-side by combining those of a 3-side and a 5-side, may we not find arcs of sides of other regular n-sides by other combinations? To take the most general case, let us form the difference of p arcs of a $2^r \cdot 3$-side and q arcs of a $2^s \cdot 5$-side; it will be the $\left(\dfrac{2^s \cdot 5 - 2^r \cdot 3}{2^s \cdot 3 \cdot 2^r \cdot 5}\right)$th of a full angle; *i.e.* it will be $(2 \cdot 5 - 2 \cdot 3)$

times the arc of one side of a $15 \cdot 2^{r+1}$-side; but this latter polygon may be constructed by the preceding problem. Hence nothing new is obtained by the new combination. Herewith, then, the round of elementary construction of regular polygons is practically completed in the four series: 2^n-sides, $3 \cdot 2^n$-sides, $5 \cdot 2^n$-sides, $15 \cdot 2^n$-sides. The profound analysis of Gauss has indeed shown that ruler and compasses will suffice to construct a (2^n+1)-side whenever (2^n+1) is a prime number; and accordingly we can construct regular 17-sides ($n=4$) and 2 5 7-sides ($n=8$); but the construction of the former is exceedingly tedious, and that of the latter is excessively so, while for still higher values of n the tedium and difficulty surpass all limit. However, in figures 94, 95 a regular 7-side and a regular 9-side are constructed once for all, empirically, but to practical perfection.

266. Problem XXIX. — *To draw a circle through two given points, tangent to a given ray.*

Hint. Consider the power of the intersection of the given ray and the ray through the points with respect to the required circle, and use Problem III.

267. Problem XXX. — *To draw a circle through a given point and tangent to two given rays.*

Hint. Find a second point on the circle and apply Problem XXIX.

268. The doctrine of perspective similarity may often be used in constructions.

A. *When one datum is a tract, the other data being angular and proportional relations.* We then construct in accordance with these latter, disregarding the first one; in the constructed figure a tract will correspond to the given tract, and on this latter we then construct the required figure similar to the one first constructed.

269. Problem XXXI. — *Given the angles and an altitude of a △, to construct it* (Fig. 163).

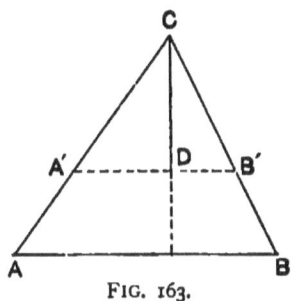

FIG. 163.

Solution. Draw any △ with the given angles; then from the proper vertex, C, lay off the given altitude as CD normal to the opposite side AB. Draw through D a ∥ to AB cutting the other side at $A'B'$. Then $A'B'C$ is the required △ (why?). The two △ ABC and $A'B'C$ are perspectively similar, C being the centre of similitude.

270. B. *When one figure is to be inscribed in another so that certain points of the one fall on certain lines of the other*, we may draw a figure in perspective, with the required figure, as to the intersection of two rays on which are to lie two points, and then from this centre of similitude construct the required figure according to the remaining conditions.

271. Problem XXXII. — *To inscribe a square in a △ with the vertices of the square on the sides of the △* (Fig. 164).

Solution. Inscribe any square HP in the △, and draw AP meeting BC at P'; then P' is a point of the required square. Complete the construction.

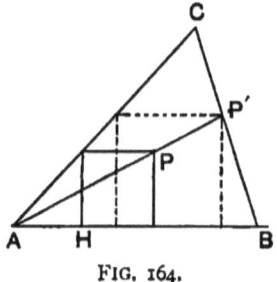

Fig. 164.

272. C. It is often required *to inscribe in a given figure a tract that shall be cut by a given point proportionally in a given ratio.* We may then assume the given point as centre of similitude, construct a figure similar to the given figure with the given ratio of similitude; then the required tract will go through a point of intersection of the two figures.

273. Problem XXXIII. — *To draw through a point I in a circle S (of radius r) a chord that shall be divided by I in the ratio $a:b$* (Fig. 165).

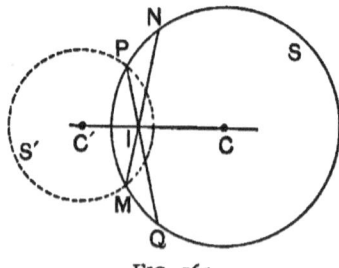

Fig. 165.

Solution. From I lay off opposite to IC the tract IC' so that $a:b::IC:IC'$. About C' as centre with radius r', such that $a:b::r:r'$, draw a circle S' meeting S at M and P. Then MN or PQ is the chord sought (why?).

CONSTRUCTIONS. 207

How will you proceed in case of outer division? The two divisions, inner and outer, may be conveniently distinguished by prefixing the sign — to the smaller term of the ratio; *i.e.* to the tract corresponding to the tract that will be wholly without the given tract after division.

*274. The following discussions might have been introduced much earlier, at Miscellaneous Applications, but for interrupting the course of thought.

We may conceive the area of a parallelogram as generated by slipping one of its sides along the other two parallel sides. Plainly, the side slipped is merely slipped or pushed, not turned at all, being kept parallel to itself (as the phrase is) throughout. Thus, suppose the tract AB slipped along the ∥ and equal tracts AD and BC; it will generate the parallelogram area $ABCD$.

Clearly, the tract may be slipped along the same parallels in either of two opposite senses, as from A to D or from D to A. The sense of the motion of the tract will be the same

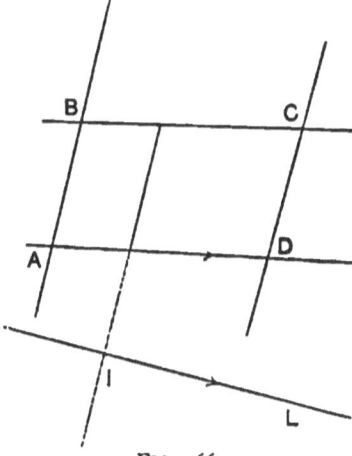

FIG. 166.

as the sense of the motion of any point of its ray, as of I, the intersection of the ray with any other ray, as with the normal ray L (Fig. 166). The two senses of I's motion may be distinguished as positive and negative; then the corresponding areas generated by the moving tract may also be distinguished as positive and negative. In summing such areas we *always regard the sense* and remember that to add resp. subtract a magnitude is the same as to subtract resp. add the *counter magnitude, i.e.* the magnitude equal in size but opposite in sense. Bearing this in mind we may now enounce:

*275. **Theorem CXXX.** — *The sum of the areas generated in simply slipping a tract round a $\triangle$ is* 0.

Data: ABC the $\triangle$, AA' the tract in its initial position (Fig. 167).

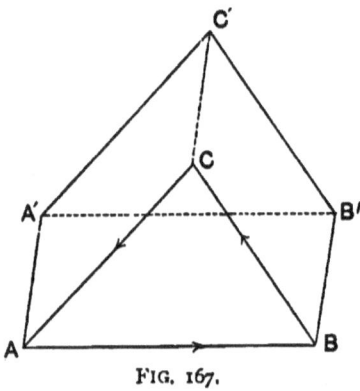

FIG. 167.

Proof. Suppose the tract to compass the $\triangle$ counter-clockwise; then if the area AB' be considered positive, the areas BC' and CA' must be considered negative (why?). But on taking away the $\triangle$ $A'B'C'$ from the whole figure AB $B'C'A'$ there is left AB'; and on taking away ABC there is left the sum of BC' and CA'; hence the areas AB' and $BC' + CA'$ are equal in size but opposite in sense; hence their sum is 0, or $AB' + BC' + CA' = 0$.   Q. E. D.

Corollary. If the $\triangle ABC$ be curvilinear instead of rectilinear, the theorem still holds.

*276. Theorem CXXXI. — *If a tract be simply pushed round any closed figure, the sum of the areas generated will be* 0 (Fig. 168).

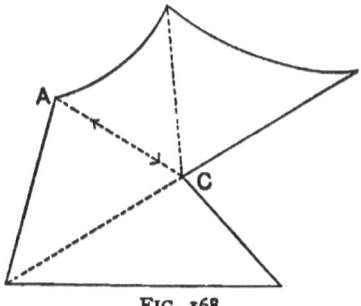

FIG. 168.

Proof. The figure may be cut up into a number of $\triangle$ rectilinear and curvilinear. The sum of areas generated in compassing each $\triangle$ is, by the foregoing theorem, 0; hence the total sum of areas generated is 0; but each dividing tract, as AC, is compassed twice, in opposite senses, from C to A and from A to C; hence the sum of areas generated along these divisions is 0; subtracting which we have left the sum of areas generated along the outer border equal to 0.   Q. E. D.

*277. Theorem CXXXII (of Pappus, A.D. 300). — *A parallelogram on one side of a $\triangle$ whose counter-vertex lies between two parallel sides of the parallelogram, equals the sum of two parallelograms, on the other sides, whose parallel sides go through the vertices of the first parallelogram.*

Proof. Let the student show from the figure, by help of Theorem CXXX that $AB' = BC' + CA'$ (Fig. 169).

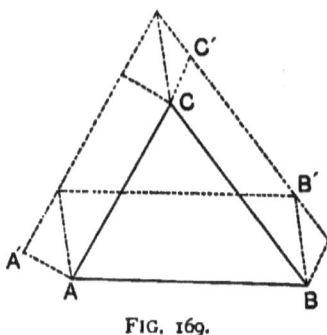

FIG. 169.

Corollary. As a special case, let the student prove the Pythagorean Theorem.

278. *Def.* The tract from a fixed point to a variable (or moving) point is called the **radius vector** of the moving point with respect to the fixed point. Thus OP is the radius vector as to O of the point P as it traces the curve C (Fig. 170).

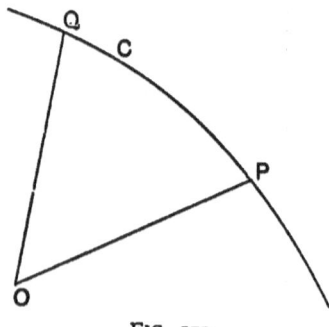

FIG. 170.

Def. The area bounded by the path of the moving point and two positions of its radius vector is said to be **generated,**

described, or *swept out* by the radius vector in passing from the initial to the final position. Thus the area POQ is swept out by r in passing from OP to position OQ.

Clearly, the same area may be described in either of two opposite senses, according as the rotation of the radius vector is clockwise or counter-clockwise, and the area must be distinguished accordingly.

We shall call areas generated clockwise negative, and areas generated counter-clockwise positive. Remembering the laws for adding and subtracting magnitudes opposite in sense, we now enounce:

279. **Theorem CXXXIII.** — *The total area generated by a radius vector whose end compasses a $\triangle$ completely is the $\triangle$ itself.*

Proof. If the point O be within or on the $\triangle$, the validity of the theorem is immediately evident. If the point O be (Fig. 171) without the $\triangle$, then the area inside is generated

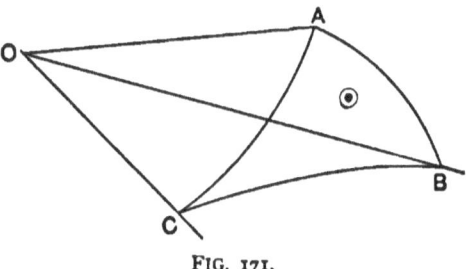

FIG. 171.

but once, while the area outside, as AOC, is generated twice, in opposite senses; once, as AOC, clockwise, once, as COA, counter-clockwise: such will always be the case, since the final and initial positions of the radius vector are the same.

Hence the outside areas annul each other, and there is left only the inside area, the △. Q. E. D.

Corollary. If the △ be curvilinear, the theorem still holds.

280. Theorem CXXXIV. — *The area described by a radius vector whose end compasses any closed figure is the area of the figure itself.*

Proof. Employ the method and reasoning of Theorem CXXXI. Conduct the proof carefully in the case of a ring and of a loop. Why do the arrows point as they do? What effect will reversing one have on the other? Imagine the ring slit through from outer to inner border (Fig. 172).

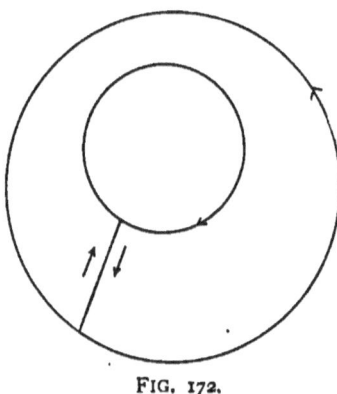

FIG. 172.

The foregoing theorems play an important rôle in Higher Mathematics.

THE TACTION PROBLEM.

281. In the following discussions certain higher concepts of Geometry, which have thus far been lightly passed over or not formed at all, become regulative and must receive graver consideration. We begin by re-defining some of them and recalling some of their already demonstrated properties.

1. The rectangle of the distances of a point from a circle along any ray through the point is called the **power** *of the point as to the circle*. The power is equal to the *square on the tangent-length* from the point to the circle when the point is without, and equal to the *square on half the shortest chord* through the point when the point is within the circle.

2. All points that have equal powers as to two circles lie on a ray called the **power-axis** of the two circles. The ray is normal to the centre-tract of the circles, of radii r and r', and divides it into segments d and d' such that $(r + r') \cdot (r - r') = (d + d')(d - d')$. The three power-axes of three circles, taken in pairs, concur in a point called the **power-centre** of the three circles.

3. Any two points P and P' on the same ray through a fixed point O are said to be in **perspective** or **perspectively similar** as to the **centre of similitude** O in the **ratio of similitude** $OP: OP'$.

4. Two figures are said to be *in perspective* or *perspectively similar* when every point of one is perspectively similar to the *corresponding* point of the other as to the same centre and in the same ratio of similitude.

5. When OP and OP' have the same sense, the perspective is *direct*, and the centre outer; when they are opposite in sense, the perspective is *counter*, and the centre inner.

6. Any two circles are perspectively similar in the ratio of their radii as to both an inner and an outer centre; namely, the points dividing the centre tract harmonically in the ratio of the radii, which are also the points of intersection of common tangents to the two circles, when such tangents there are.

282. Theorem CXXXV. — Lemma. — *If two figures are similar to a third, they are similar to each other.*

The easy proof is left to the student.

283. Theorem CXXXVI. — *If one figure is in perspective with each of two, these latter are in perspective with each other, and the three centres of similitude are* **collinear.**

Proof. Let P, Q, R be any three points in the first figure, P', Q', R' and P'', Q'', R'' the corresponding points in the other figures. Then PQR and $P''Q''R''$ (Fig. 173) are similar (why?), and PP'', QQ'', RR'' meet in a point, O'

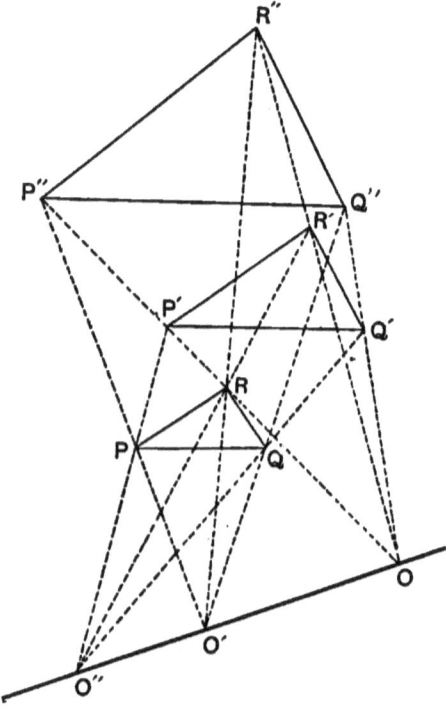

FIG. 173.

(why?), as to which they are in perspective (why?). Likewise PQR and $P'Q'R'$ are similar, and PP', QQ', RR' meet in a point O''.

Now let PQ, $P'Q'$, $P''Q''$ cut $O'O''$ at S, S', S'', and draw SR, SR', SR''. Then QRS, $Q'R'S'$, $Q''R''S''$ are all similar (why?), S and S' are in perspective as to O'', S and S'' are in perspective as to O', and hence S' and S'' are in perspective as to O. Hence O is on the ray $O'O''$ (why?). Q.E.D.

Corollary. Show that the three centres of similitude are either *all outer* or else *one outer* and *two inner*.

Def. If a point bisect every chord of a figure drawn through the point, it is called the **centre** of the figure, and the figure itself is said to be **centric**.

A central ray, and often a central chord, of the figure is called a **diameter**.

284. **Theorem CXXXVII.** — *If two similar centric figures be in perspective as to one point, they are also in perspective as to a second point* (Fig. 174).

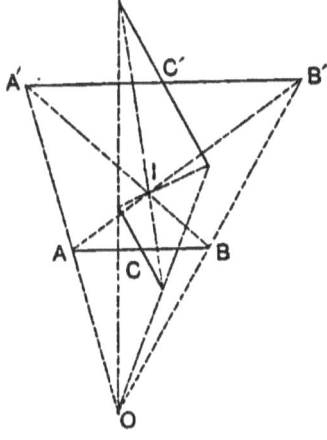

FIG. 174.

Proof. Draw two diameters of the one figure and the two corresponding chords of the other; these latter will also be diameters (why?), and the centres will correspond. Join the ends of these diameters *crosswise;* that is, the end of one with the non-corresponding end of the other. The intersection of these two cross-joins, I, is a second centre of similitude (why?). Q. E. D.

Corollary 1. Of these two centres of similitude, the one is outer, the other inner; and they divide the centre tract CC' harmonically.

Corollary 2. If three similar centric figures be in perspective they have six centres of similitude, and of these the three outer are collinear, as are also any one outer and the two other inner.

Def. A ray on which lie three centres of similitude is called an **axis of similitude.**

Corollary. There are four such axes, one outer and three inner.

The central figure with which we have especially to do is the *circle.*

285. *Def.* When the rectangle of the distances from a fixed point O of two points, P and P', on the same ray through O, is constant, the two points are said to be **inverse,** or **in inversion,** with respect to O as **centre of inversion.**

Def. A circle about the centre of inversion, with the side of the square equal to the rectangle of the distances for radius, is called the **circle of inversion,** and its radius the **radius of inversion.**

Def. If while one of the inverse points as P describes a curve C the other describes a curve C', then C and C' are said to be **inverse** or **in inversion** with respect to O.

TH. CXXXVIII.]   *THE TACTION PROBLEM.*   217

Def. Let a ray through a centre of similitude of two circles cut each in a pair of correspondent points P and Q, P' and Q'; then each point of each pair has a correspondent or homologous point in the other pair, as P and P', Q and Q'; also each point of each pair has a **non**-correspondent, or **contra**-correspondent, or **anti**-homologous point in the other pair, as P and Q', Q and P'.

286. **Theorem CXXXVIII.** — *Anti-homologous points of two circles are inverse with respect to the centre of similitude of the circles* (Fig. 175).

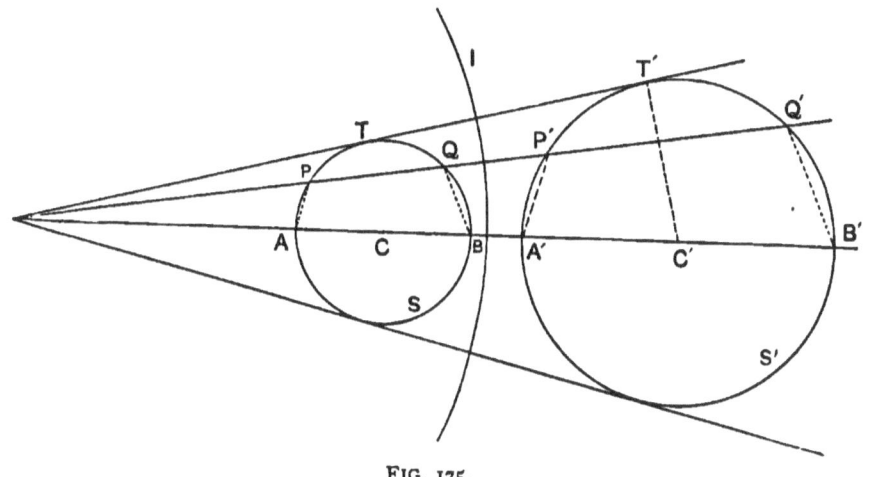

FIG. 175.

Proof. Let O be the centre of similitude of the circles, CC' the centre ray.

Then $\angle OAP = \angle O'A'P'$ (why?) and $\angle OAP = \angle BQP$ (why?); hence $\angle BQP = \angle BA'P'$; hence $\angle BA'P'$ and $\angle BQP$ are supplemental; hence $BA'P'Q$ is an encyclic quadrangle; hence $OQ \cdot OP' = OB \cdot OA'$. Now the points O, B, A' are fixed; hence the rectangle $OB \cdot OA'$ is constant; hence Q and P' are inverse as to O. Similarly prove that P and Q' are inverse.   Q. E. D.

Corollary. $OA \cdot OB' = OT \cdot OT'$; hence the radius of inversion about O is the geometric mean of OT and OT', the tangent-lengths from the centre of similitude to the circles.

287. **Theorem CXXXIV**[a]. — *The inverse of a circle is itself a circle.*

Data: In the figure let O be the centre of inversion, S the circle, I the circle of inversion with radius r.

Proof. Draw OT tangent to S and construct OT' so that $OT : r :: r : OT'$. Draw a normal to OT at T' meeting OC

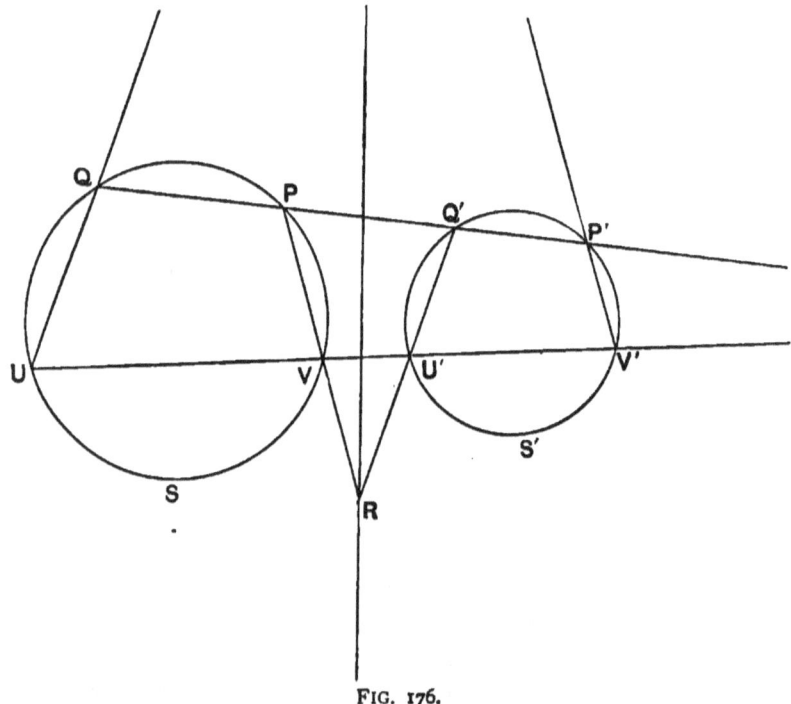

FIG. 176.

at C'; with radius $C'T'$ draw a circle. It is the inverse sought. For it is the circle S' of the preceding theorem (why?), in which P and Q', Q and P' were inverse as to O.

288. Theorem CXXXV[a]. — *The transverse joins (chords) of two pairs of anti-homologous points of two circles meet on the power-axis of the circles.*

Data: P and Q', V and U', two pairs of anti-homologous points in S and S'; PV and $Q'U'$, their transverse joins (chords) (Fig. 176).

Proof. The quadrangle $PVU'Q'$ is encyclic (why?). Let the student complete the proof.

Def. If a circle touch two other circles, the ray through the points of touch is called the *chord* (or, better, the **ray**) **of Contact**.

289. Theorem CXXXVI[a]. — *The ray of contact of two circles with a third goes through a centre of similitude of the two circles* (Fig. 177).

Proof. For a point of contact of two circles is a centre of similitude of the two (why?) ; hence the ray of contact goes through two centres of similitudes ; hence it is an axis of similitude and goes through a third centre of similitude ; namely, of the two circles (why?). Q. E. D.

Corollary. When the two circles are touched *similarly*, both innerly or both outerly, the ray of contact goes through the *outer* centre of similitude of the two ; when they are touched *dissimilarly*, one innerly the other outerly, the ray of contact goes through the *inner* centre of similitude of the two (why?).

290. *Def.* The ray through one of two inverse points normal to their junction-ray is called the **polar** of the other

220   GEOMETRY.   [TH. CXXXVII^a.

point, and this latter point is called the **pole** of the polar, *with respect to the circle of inversion*. This circle we may call the *circle of reference*, or the *referee-circle*, or simply the **referee**. Note carefully that pole and polar have no meaning except with respect to some referee.

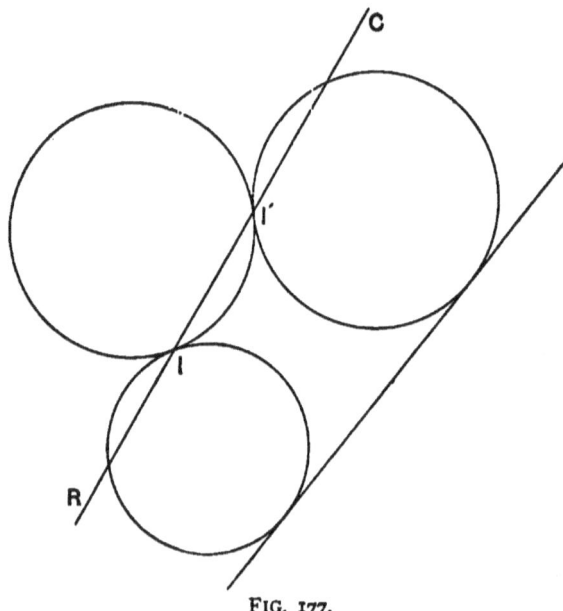

FIG. 177.

291. Theorem CXXXVII^a. — *If of two points the first is on the polar of the second, then the second is on the polar of the first.*

Data: S the circle, P the first point on the polar, L, of the second point Q (Fig. 178).

Proof. Draw the centre-ray OQ cutting L at Q', then Q' is the inverse of Q (why?) ; also, let P' be the inverse of P. Then the quadrangle $PP'QQ'$ is encyclic (why?) ; also the $\angle Q'$ is a right angle (why?); hence so is the $\angle P'$ (why?); hence $P'Q$ is the polar of P (why?), *i.e.* the polar of P goes through Q. Q.E.D.

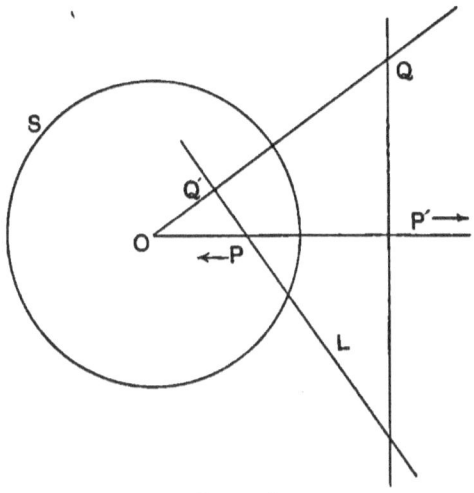

FIG. 178.

Corollary 1. The poles of all rays through P lie on the polar of P; in other words, as a polar turns about a point, its pole glides along the polar of that point.

Corollary 2. The polars of all points on a ray pass through the pole of that ray; in other words, as a pole glides along a ray, its polar turns about the pole at that ray.

Scholium. By definition the rectangle $OP \cdot OP'$ is constant in area; hence as P moves in towards O, P' moves out, and with it the polar of P; as P falls on O, P' and the polar of P through it move out and vanish in infinity; as P moves out from O leftward, P' and the polar reappear in infinity on the left, approaching S; as P reaches S, so does P', and the polar becomes a tangent. Hence we may define the tangent as *a polar whose pole is on it* (the polar).

292. Theorem CXXXVIII'. — *Tangents at the end of a chord meet on the polar of the chord* (Fig. 179).

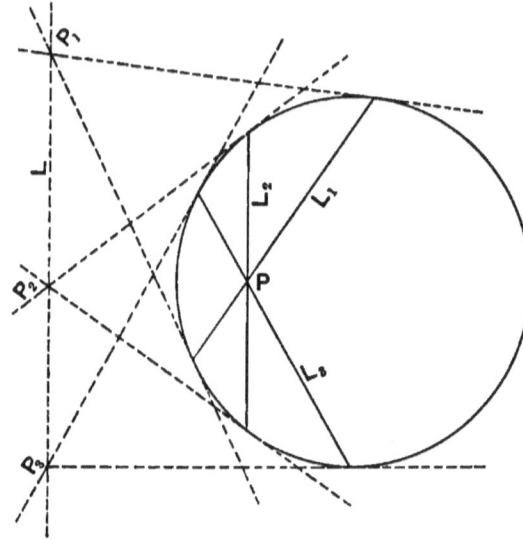

FIG. 179.

Proof. For the chord goes through the poles of the tangents (where?); hence the tangents go through the pole of the chord (why?).

Corollary. Tangents at the end of a chord through a point meet on the polar of the point.

Hence we may define the polar of a point as *the locus of the intersection of the pair of tangents at the ends of any chord through the points.*

Exercise. Show how to construct the polar of a point without, within, or upon the circle of reference.

Def. Two points, each on the polar of the other, and two polars, each through the pole of the other, are called **conjugate**.

We are now prepared to attack

THE TACTION PROBLEM.

To draw a circle tangent to three given circles.

293. Lemma A. — *The power-centre of three circles is a centre of similitude of two circles each tangent to each of the three.*

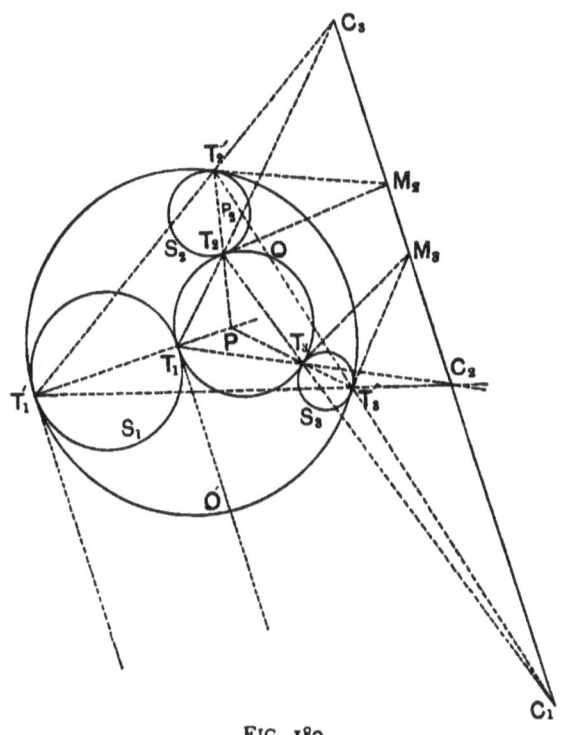

FIG. 180.

Data : S_1, S_2, S_3, the three circles, O and O' two circles touching them, O outerly, O' innerly ; T_1, T_2, T_3, T'_1, T'_2, T'_3 the points of touch.

Proof. Draw the rays of contact, $T_1T'_1$, $T_2T'_2$, $T_3T'_3$.
The first goes through two points of dissimilar contact of

S with O and O'; hence it goes through the inner centre of similitude of O and O'; the like may be said of $T_2T'_2$ and $T_3T'_3$; hence these rays concur in P, the inner centre of similitude of O and O'.

Hence T_1 and T'_1, T_2 and T'_2, T_3 and T'_3 are three pairs of anti-homologous points on O and O' with respect to P; hence $\quad PT_1 \cdot PT'_1 = PT_2 \cdot PT'_2 = PT_3 \cdot PT'_3$;

that is, $\quad P$ is the Power-centre of S_1, S_2, S_3. Q. E. D.

294. Lemma B. — *An axis of similitude of three circles is a power-axis of two circles tangent each to each of the three.*

Data: The same as before.

Proof. The transverse joins T_1T_2 and $T'_1T'_2$ meet on the power-axis of O and O' (why?); so too the transversals T_2T_3 and $T'_2T'_3$, T_3T_1 and $T_3T'_1$. But T_1T_2 and $T'_1T'_2$ are rays of contact of O with S_1S_2 and of O' with S_1S_2; hence they meet in the (outer) centre of similitude of S_1 and S_2; similarly for the pairs T_2T_3 and $T'_2T'_3$, T_3T_1 and $T'_3T'_1$. Hence the outer axis of similitude of S_1, S_2, S_3 is the power-axis of O and O'.   Q. E. D.

295. Lemma C. — *The ray of contact of each circle with the two circles goes through the pole as to the circle of an axis of similitude of the three circles.*

Data: The same as before.

Proof. The tangents M_2T_2 and $M_2T'_2$ are equal; hence the point M_2 is on the power-axis of O and O', *i.e.* on the (outer) axis of similitude of S_1, S_2, S_3. So, too, for M_3 and M_1, which latter in the figure lies at infinity. But M_2 is the pole as to S_2 of the contact-ray $T_2T'_2$ (why?); hence the pole of the contact-ray lies on the axis of similitude; hence the pole of the axis of similitude lies on the contact-ray, or the contact-ray goes through the pole of the axis of similitude.   Q. E. D.

THE TACTION PROBLEM.

296. Accordingly, we know two points of each contact-ray, namely, the *power-centre* of the three circles and the *pole* of an axis of similitude as to each circle. We have then this rule of construction :

Solution. (1) Find the power-centre and an axis of similitude of the three circles; (2) find the pole of this axis as to each of the circles; from the power-centre draw three rays through the three poles: they cut the three circles in three pairs of points, namely, the points of tangency of two required circles.

Thus it appears that each axis of similitude yields in general two tangent circles; and there are four such axes; hence there are in general *eight* tangent circles.

The kind of tangency is determined by the axis of similitude : if this be outer, then each of the two circles touches all three similarly, one outerly, the other innerly; if the axis be inner, but drawn through the outer centre (say) of S_1 and S_2, then one of the circles will touch S_1 and S_2 outerly, but S_3 innerly, while the other will touch S_1 and S_2 innerly, but S_3 outerly.

297. This classic problem, in which the elementary geometry of the circle seems to culminate, was proposed and solved by Apollonius of Pergæ, A.D. 200. His solution was indirect, reducing the problem to ever simpler and simpler problems. It was lost for centuries, but was restored by Vieta. The direct solution similar to the foregoing was first given by Gergonne (1813). The analogous problem for space, namely, to construct a sphere tangent to four given spheres, was first solved by Fermat (1601-1665). The foregoing construction is immediately applicable to this problem, on changing 3 into 4 and ray into plane.

It is important that the student actually carry out the preceding solution.

METRIC GEOMETRY.

298. Thus far our treatment of the subject of Geometry has been strictly geometrical; we have at no point invoked the aid of number, Arithmetic, or Algebra in demonstration, so that if these sciences should suddenly vanish from cognition the structure of our geometric knowledge would remain wholly unimpaired. Nevertheless, in the Art of Geometry, in the practical application of the science to quantitative problems, it becomes highly important to express linear, angular, and areal magnitudes through numbers or at least numerically, and to apply to such expressions the laws of numerical calculus. Such is the subject of the following sections.

299. *Def.* A geometric magnitude (tract, angle, area) that may be regarded as the sum, or that equals the sum, of m equal geometric magnitudes (of the same kind) is called a

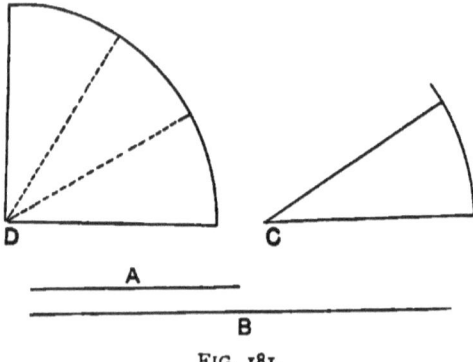

FIG. 181.

multiple (more precisely, the **m-fold**) of one of those equal magnitudes. Thus B is the double of A, D the triple of C, F the four-fold of E (Figs. 181, 182).

Def. Any one of m equal geometrical magnitudes is called an **mth part**, or simply an **mth**, of the sum or of the equal of the sum of these m magnitudes.

Thus A is a half of B, C is a third of D, E is a fourth (part) of F.

The symbol for the m-fold of a magnitude is formed by prefixing m to the symbol for the magnitude. Thus, $2A$, $3C$, $4\square$, mM. The symbol for an mth part of a magnitude is commonly formed by writing m below the magnitude and separating the two by a horizontal or oblique bar: thus,

$$\frac{B}{2}, \frac{D}{3}, \square/4, Q/m.$$

300. **Theorem CXXXIX.** — *The p-fold of the mth part of a magnitude equals the mth part of the p-fold of the magnitude.*

Proof. Let Q be any magnitude (tract, angle, area). By definition there are m mth parts of it. In its p-fold each such part will be present p times; hence there will be pm mth parts in the p-fold of Q.

Also by definition the mth part of this p-fold taken m times in summation must yield the whole. Now, however, if we take p of the mth parts of Q and take them m times in summation, we shall get a whole consisting of mpmth parts. But it is a fundamental law of counting, called the Commutative Law of Multiplication, that to count m p times yields the same number as to count p m times. Hence this whole is equal to the p-fold of Q; and its mth part consists of p mth parts of Q, or is the p-fold of the mth part of Q; *i.e.* the p-fold of the mth part of Q equals the mth part of the p-fold of Q. Q.E.D.

Scholium. The *p*-fold of the *m*th part, or the *m*th part of the *p*-fold, of Q is commonly written

$$\frac{pQ}{m}, \text{ or } \frac{p}{m} \cdot Q \text{ or } p \cdot \frac{Q}{m}.$$

The expression $\frac{p}{m}$, or p/m is called a **fraction**, p and m its *terms*, p the *numerator*, m the *denominator*. We have just learned what it means.

301. If now we conceive any whole, w, as the sum of m equal parts, each equal to u, we may call u the **unit magnitude** or **magnitudinal unit**. Thus one yard is a linear, one degree an angular, one acre an areal, unit. There may be several other magnitudes, the *p*-fold, *q*-fold, *x*-fold of this same unit u. Then m, p, q, x are called the **metric numbers** of these magnitudes.

302. It may happen that a magnitude may not be composable out of equal units u; it may not be a multiple of the unit-magnitude u, but may be greater than the *p*-fold of u and less than the $(p+1)$-fold of u. Thus a circle is more than triple, yet less than quadruple, its diameter. In such cases it may be possible to find some smaller unit of which the unit u is the *m*-fold, and the other magnitude (say) the *q*-fold. Thus the table may be more than 3 feet and less than 4 feet long; but on changing the linear unit from the foot to the inch, the twelfth of a foot, we may find that the length in question is precisely the 40-fold of the new unit — the table is precisely 40 inches long. Then 40 is the metric number of the table-length in inches and the fraction $\frac{40}{12}$ is the metric number of the same length in feet, which means that the sum of 40 12th parts of a foot is the length of the table.

303. Often, however, in fact generally, it will be impossible to find any unit-magnitude so small that its m-fold shall be the one magnitude and its p-fold the other; and this impossibility may be objective, not subjective — it may inhere in the nature of the case and not arise from some defect of our own powers of measurement or calculation. Thus, there is no unit-length, however small, out of which may be composed both the side and the diagonal of a square; there is no length so small that the side shall be its m-fold and the diagonal its q-fold. This important fact may be established thus:

304. Lemma. — *If each of two tracts is a multiple of the same tract, the difference of the two is a multiple of the same tract.*

Proof. Let G be the greater and L the less of the two tracts. Then we have $G = pt$ and $L = q \cdot t$; the difference of these two is $(p - q)t$, and this is a multiple of t, since the difference of two integers, p and q, is itself an integer, $p - q$.

305. Theorem CXL. — *The side and diagonal of a square are incommensurable* (Fig. 182).

Proof. Let $A_1B_1 = s_1$ be the side, and $A_1A_2 = d_1$ be the diagonal of a square. On A_1A_2 lay off $A_1B_2 = s_1$; then A_2B_2 is the difference of the side and diagonal and is therefore a multiple of any tract of which s_1 and d_1 are multiples; call it s_2. Draw B_2A_3 normal to A_1A_2; then $B_1A_3 = B_2A_3$ (why?) $= A_2B_2$ (why?). Hence A_2A_3 is the diagonal, d_2, of a square whose side is A_2B_2 or s_2. Also d_2 is a multiple of any tract of which s_1 and s_2 are multiples. Hence we have a new square with side s_2 and diagonal d_2, both multiples of any tract of which s_1 and d_1 are multiples. Also the new side and diagonal are respectively less than half of the old side and diagonal. By repeating this process we obtain a third square with side and diagonal less than half the side and diagonal of the second, less than one fourth those of

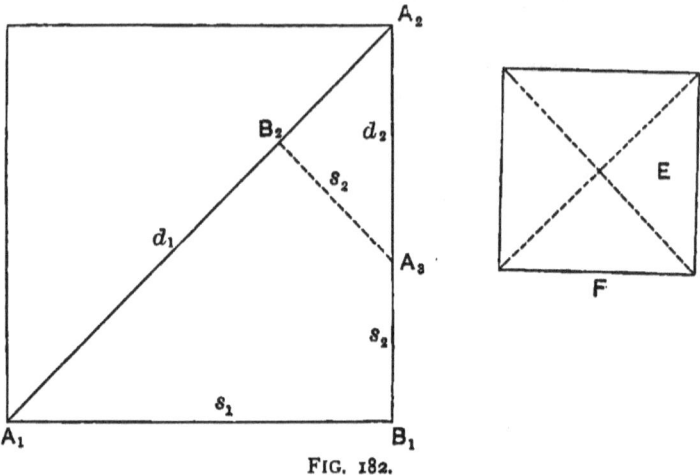

FIG. 182.

the first. By such constructions we shall obtain a square with side and diagonal less than $\left(\dfrac{1}{2^n}\right)$th of the side and diagonal of the original square; *i.e.*

$$s_n < \frac{s_1}{2^n},\ d_n < \frac{d_1}{2^n}.$$

By making n large enough we may make s_n and d_n as small as we please, *small at will,* — we may in fact sink them below any assigned degree of parvitude. Meanwhile they must remain multiples of any tract of which s_1 and d_1 are multiples; but they can be multiples only of a tract smaller than themselves, manifestly; hence any tract of which both s_1 and d_1 are multiples must be *smaller than a tract as small as we please.* But there is no such tract, self-evidently. Hence there is no tract of which both diagonal and side of a square are multiples. Q. E. D.

Magnitudes that are thus not multiples of the same magnitude are said to have no common measure, or to be **incommensurable**.

306. Now suppose the side s cut up into some very large number of equal parts, as q; then the diagonal d will not be the sum of any number of these parts but will be more than the sum of p parts and less than the sum of $(p+1)$ parts;

that is, $\quad \dfrac{p}{q} \cdot s < d < \dfrac{p+1}{q} \cdot s.$

Here we may make q as large as we please; hence, if we take s for our linear unit, we may shut in the metric number of d between two fractions, $\dfrac{p}{q}$ and $\dfrac{p+1}{q}$, that differ by $\dfrac{1}{q}$, that is, by a fraction *small at will*, while the *metric number* of d differs from each by less than $\dfrac{1}{q}$. In this last sentence we have subreptively assumed that d *has some metric number* when referred to s as unit-length. But we shall not build anything on this assumption at present. See Art. 256.

307. We now pass to the metric numbers of area. We agree once for all that the **square on a linear unit** shall be an **areal unit**. The *metric number* of an area will then be the number of areal units of which it is composed, or its equal is composed.

308. **Theorem CXLI.** — *The metric number of a rectangle is the product of the metric numbers of its dimensions.* Two cases arise:

1. *When the dimensions are commensurable* (Fig. 183).

Let a and b be the dimensions of a rectangle. Choose any unit of which a and b are both multiples, as u, so that

232   GEOMETRY.   [TH. CXLI.

$a = p \cdot u$, $b = q \cdot u$. Then we may cut up a into p parts, and b into q parts equal to u. Through the points of division draw ∥s to the sides. Then the whole rectangle will be cut

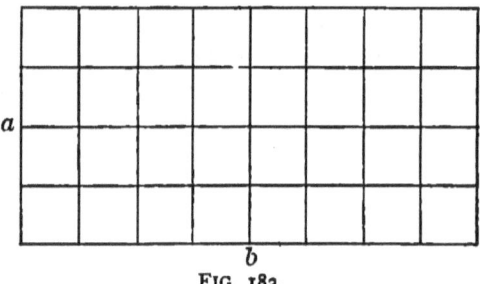

FIG. 183.

up into pq squares each on the side u (why?); hence the rectangle will be the sum of these areal units; hence the metric number is the product pq. Q.E.D.

Now suppose we choose some larger areal unit, as the square not on u but on ru. In this square there are, by the foregoing, rr units; that is, it is the rr-fold of the small unit u^2; or the small unit is the $\left(\frac{1}{rr}\right)$th part of the large areal unit; hence the rectangle, which is the pq-fold of the small unit, is the $\left(\frac{pq}{rr}\right)$th of the large areal unit. But the sides were respectively the $\left(\frac{p}{r}\right)$th and the $\left(\frac{q}{r}\right)$th of the small linear unit u; and the product of these two fractions is, *by the laws for multiplying fractions*, $\frac{pq}{rr}$. Hence, *if we call fractions numbers*, we must have the metric number of the area is $\frac{pq}{rr}$, or is the product of the metric numbers of the dimensions. Q.E.D.

2. Now suppose the two dimensions **incommensurable** with the linear unit u. Then a will be $> pu$ and $< (p+1)u$;

while b will be $> q \cdot u$ and $< (q+1)u$. Then plainly there are pq squares on u in the rectangle, with some remainder, but not $(pq + p + q + 1)$ squares; or

$$pq \cdot u^2 < ab < (pq + p + q + 1)u^2.$$

Hence, *if the rectangle ab has any metric number at all*, the same must lie between the values pq and $(p+1) \cdot (q+1)$; *i.e.* it must lie between the product of the metric numbers too small and the metric numbers too great for the dimensions. Now u was very small, and hence p and q very large. Take a new unit $U^2 = rr \cdot u^2$, whose side is the rth multiple of the side of the same square. Then the metric numbers of the dimensions a and b will be $> \frac{p}{r}$ and $\frac{q}{r}$ but $< \frac{p+1}{r}$ and $\frac{q+1}{r}$, and the metric number of the area, *if it have a metric number*, will be $> \frac{pq}{rr}$ but $< \frac{(p+1)(q+1)}{rr}$; *i.e.* it will lie between these two fractions and differ from each by less than $\frac{p+q+1}{rr}$. Now with any fixed unit length, as 1, we may find two fractions, $\frac{p}{r}$ and $\frac{q}{r}$, that differ from the metric numbers of a and b (if they have any) by less than $\frac{1}{r}$, and by taking r even greater and greater we may approach our fractions $\frac{p}{r}$ and $\frac{q}{r}$ close at will to the metric numbers of a and b. Each of these fractions $\frac{p}{r}$ and $\frac{q}{r}$ meanwhile remains less than some assignable whole number; so, too, does their sum $\frac{p+q}{r}$, and so does $\frac{p+q+1}{r}$. Now the difference of the fractions $\frac{pq}{rr}$ and $\frac{(p+1)(q+1)}{rr}$ is the rth part of $\frac{p+q+1}{r}$, and by making r *large at will* we may make this *r*th part *small at will.* Hence the two fractions may be brought as close together in value as we please, while between them lies always the metric number of the area, and also between them lies always the product of the metric numbers of the dimensions. These two *numerics*, then, the metric number of the area and the product of the metric numbers of the dimensions, cannot differ by any assignable value however small, since they both lie between two values which may be made to differ by less than any assigned value however small. Hence we conclude (1st) that these two numerics are **Definites**, since the bounds between which each lies, and which close down together upon each other, are at every stage perfectly definite, and (2d) that they are absolutely the same in value.

309. It is a matter not of logical compulsion but of convenient choice to call this **Definite** a *number* or at least a *numeric*. Since it is not expressible as a fraction, still less an integer, it is commonly called an **Irrational**. The laws of operation on the algebraic symbols of such Irrationals as well as Fractions are not matters of logical proof, but of allowable assumption. It is convenient to assume for them, arbitrarily to impose upon them, the same laws of operation that are found empirically to hold for positive integers, or numbers obtained by counting. This fact is sometimes called the **Principle of the Permanence of the Formal Laws of Operation** (Hankel). Further discussion of the subject belongs to Algebra and would be out of place here.

310. Knowing that the metric number of a rectangle is the product of the metric numbers of its dimensions, we now declare at once that

Theorem CXLII. — *The metric number of a parallelogram is the product of the metric numbers of its dimensions.*

Theorem CXLIII. — *The metric number of a △ is half the product of the metric numbers of its dimensions.*

Theorem CXLIV. — *The metric number of a trapezoid is half the product of the metric numbers of its altitude and the sum of its ∥ bases.*

In a word, all the theorems that declare relations among areas may now be translated into theorems that declare like relations among the metric numbers of areas. This easy exercise is left for the student.

---

311. We have thus far treated proportion strictly geometrically. We have written off the symbolism

$$a:b::c:d,$$

when a, b, c, d, were signs for tracts, but when asked what we meant by it our only reply was, we mean that the rectangle ad equals the rectangle bc. This reply was perfect and complete. Now, however, if instead of the tracts we put p, q, r, s, as the metric numbers of the tracts, we may still write as before

$$p:q::r:s,$$

and answer the question what this means, by saying it means that the *product* ps equals the *product* qr, for we have just proved this equality. This answer is also perfect and complete. However, it is not the *only* possible answer. For we might say we mean that the fraction $\frac{p}{q}$ equals the fraction $\frac{r}{s}$; this would also be correct. For if $ps = qr$, then on dividing both sides by qs we get $\frac{p}{q} = \frac{r}{s}$, and conversely, if $\frac{p}{q} = \frac{r}{s}$, then on multiplying both sides by qs we get $ps = qr$. Accordingly these two answers are equally adequate and involve each other. But we could not make any such second answer to the question, what do we mean by the proportion

$$a:b::c:d?$$

For we cannot attach any meaning to the symbolism $\frac{a}{b}$ when a and b are tracts, nor can we tell, at least at present, what we mean by dividing one tract by another. We may indeed write $\frac{a}{b} = \frac{c}{d}$, and answer the inquiry as to our meaning by saying we mean the rectangle ad equals the rectangle bc; but we cannot *deduce* the relation *rectangle ad = rectangle bc* from the symbolism $\frac{a}{b} = \frac{c}{d}$ by multiplying through by the rectangle bd, for we do not attach any meaning to the phrase " multiplying by a rectangle."

312. The state of the case then is this:

All the proportions among tracts in Geometry may be supplaced by corresponding proportions among the metric numbers of those tracts; in these latter proportions we may supplace the colon by the division —, or quotient —, or fraction-mark, and the double colon by the equality-mark. The ratio of two tracts we did not attempt, and did not need, to define geometrically; but we now define the corresponding *ratio of the metric numbers* of those tracts as the quotient of the one metric number divided by the other; and a **proportion** among these metric numbers of tracts we may define as an **equality of ratios**.

313. We may now boldly apply the ordinary laws of algebraic equations to any geometric proportion, understanding by its terms *not* the tracts themselves, but the metric numbers of the tracts. The result will be some relation among the metric numbers of tracts. If desirable, we may

at once translate this relation back into pure Geometry by substituting for the metric numbers of the tracts the tracts themselves. But it will not always be possible to interpret geometrically the result of this substitution. An illustration will make this clear.

314. Let p, r, s, t, u be the metric numbers of the tracts on which they are written in the $\triangle ABC$, right-angled at B with BD normal to AC (Fig. 184). Then $s^2 = tu$ (why?),

and $\quad s^4 = t^2 u^2 = (r^2 - s^2)(p^2 - s^2) = p^2 r^2 - p^2 s^2 + s^4 - r^2 s^2$.

Whence $\quad p^2 r^2 = p^2 s^2 + r^2 s^2$,

whence $\quad \dfrac{1}{s^2} = \dfrac{1}{p^2} + \dfrac{1}{r^2}.$

This beautiful and important relation may be stated thus:

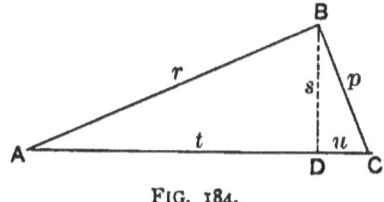

FIG. 184.

Theorem CXLIII^a. — *The reciprocal of the squared metric number of the altitude to the hypotenuse of a right triangle equals the sum of the reciprocals of the squared metric numbers of the sides.*

So stated the meaning is intelligible and unmistakable.

But if now we write for s, r, p the tracts themselves, namely,

$$\dfrac{1}{\overline{BD}^2} = \dfrac{1}{\overline{AB}^2} + \dfrac{1}{\overline{BC}^2},$$

then we may indeed understand this relation *algebraically* precisely as before, meaning by the signs $\overline{BD}^2, \overline{AB}^2, \overline{BC}^2$

the squared metric numbers of the tracts BD, AB, BC; but we cannot attach any *geometric* meaning to the equation, for we cannot tell what we mean by the reciprocal of a geometric square.

315. The next illustration is still more interesting and important (Fig. 185).

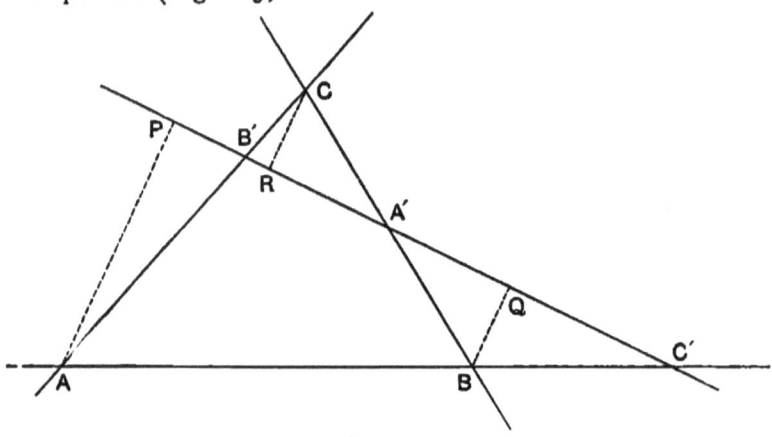

FIG. 185.

Let ABC be any $\triangle$, L any ray cutting the sides at A', B', C'; from A, B, C drop normals on L meeting it at P, Q, R.

Then by similar $\triangle$ we have
$$AC' : BC' : : AP : BQ,$$
$$BA' : CA' : : BQ : CR,$$
$$CB' : AB' : : CR : AP.$$

If now we understand by these biliterals not the tracts themselves, but the metric numbers of the tracts, the foregoing proportions will still hold and may be read as equations and written thus:
$$\frac{AC'}{BC'} = \frac{AP}{BQ}, \quad \frac{BA'}{CA'} = \frac{BQ}{CR}, \quad \frac{CB'}{AB'} = \frac{CR}{AP}.$$

where the sides of the equations are ordinary fractions. On multiplying them together there results

$$\frac{AC' \cdot BA' \cdot CB'}{BC' \cdot CA' \cdot AB'} = \text{unity}.$$

Inasmuch as AC' and BC' are reckoned oppositely as are also BA' and CA', BC' and AB', it is common and convenient to write -1 instead of 1, thus:

$$\frac{AC' \cdot BA' \cdot CB'}{BC' \cdot CA' \cdot AB'} = -1.$$

This is the celebrated proposition of *Menelaos*:

Theorem CXLIV[a]. — *The continued product of the ratios in which a ray cut the sides of a $\triangle$ is* -1.

It states the condition necessary and sufficient that three points on the sides of a $\triangle$ shall be collinear, and its meaning is perfectly clear so long as we mean by AC', etc., not the tracts but the metric numbers of the tracts. But in order to interpret it geometrically, AC', etc., standing for the tracts themselves, it would be necessary to define precisely a higher notion, namely, that of the *volume of the cuboid of three tracts*, and this would require us to pass out of our plane into tri-dimensional space.

316. Still another illustration is found in a proposition the logical complement of the preceding (Fig. 186).

Let any three rays through the vertices of a $\triangle$ concur in O, and let normals from O meet the sides of the $\triangle$ ABC at the points A', B', C'.

Draw OA, OB, OC, and form the pairs of ratios $OA' : OB$; $OA' : OC$; $OB' : OC$; $OB' : OA$; $OC' : OA$; $OC' : OB$. Regarding them as ratios not of tracts but of the metric numbers of tracts, we may treat them as fractions and form the product of the first in each of the three couplets, and also of the second in each couplet; these products are evidently equal. Now the fraction $\dfrac{OB'}{OA}$ depends for its value solely on the angle α; hence it is called a **function** of

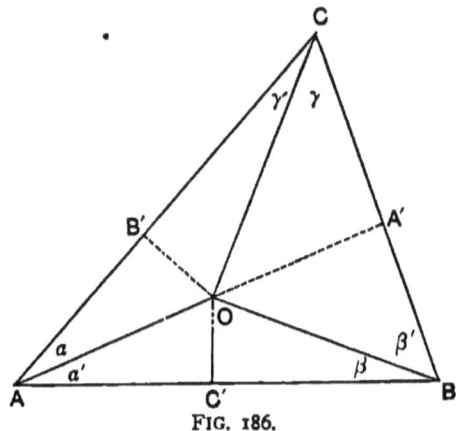

FIG. 186.

the angle α, namely, the **sine** of the angle α, and so for the others. Hence

$$\frac{\text{sine of } \alpha}{\text{sine of } \alpha'} \cdot \frac{\text{sine of } \beta}{\text{sine of } \beta'} \cdot \frac{\text{sine of } \gamma}{\text{sine of } \gamma'} = 1 \,; \text{ or,}$$

Theorem CXLV. — *The continued product of the ratios of the sines of the angles into which three concurrent rays through the vertices of a $\triangle$ divide the angles of a $\triangle$ is 1.*

The converse of this theorem is easily established, which accordingly supplies a test of the concurrence of three rays through the three vertices of a $\triangle$. Its meaning is perfectly precise and unmistakable so long as not the tracts but the metric numbers of the tracts are signified by OA, etc.; otherwise we are not in position to prove it nor to interpret the symbolism expressing it.

METRIC GEOMETRY.   241

317. The notion of *sine of an angle*, introduced for simplicity in the foregoing article, is of the highest importance for all following geometrical study. But perhaps a more fundamental notion is that of cosine, which we may define thus:

Def. The *ratio of the projection of a tract on a ray to the tract itself* is called the **cosine** of the angle between the ray and the tract (Fig. 187).

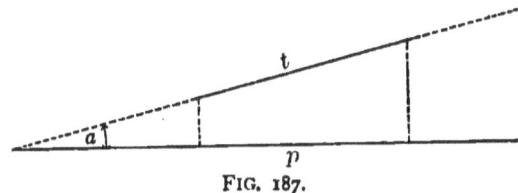

FIG. 187.

Thus the ratio of the projection p of the tract t to the tract itself is the cosine of the angle α between them; or $p : t = \underline{\alpha}$, as we may write **cosine of** α, which is commonly abbreviated into **cos** α.

318. It is plain that the projections of t on parallel rays are all equal; hence we may suppose the ray of projection drawn through the beginning of t, as in Fig. 188. Then as t turns about o and α changes its value, the projection p of t will also change. Thus: for

$\alpha =$ 0,   60°,   90°,   120°,   180°,   270°,   360°,   420°,   ⋯

$\underline{\alpha} =$ 1,   ½,   0,   −½,   −1,   0,   1,   ½,   ⋯

319. In the 2d and 3d quadrants the projection p is reckoned leftward; it is opposite in sense to the projection when α is in the 1st or 4th quadrants, and accordingly the cosine is marked −. When α increases by 360° (or 2 π, see Art. 336) from any value, the revolving tract resumes its original position, the projection resumes its original value, and so too does the cosine; hence

$$\cos(360° + \alpha) = \cos(2\pi + \alpha) = \cos\alpha;$$

that is, *the cosine is not changed by increasing* (or decreasing) *the angle by a round angle.*

Hence, plainly, $\cos(\alpha \pm 2n\pi) = \cos\alpha.$

Hence the cosine is called a **periodic function*** of the angle, the **period** being 2π, that is, a *round angle* (see Art. 336).

320. If t be turned through any angle α from the position OA, its projection p is the same whether the turning be clockwise or counter-clockwise; that is, the projection is the same whether the angle be negative or positive; hence, too, the cosine is the same.

That is, $\cos(-\alpha) = \cos\alpha;$

that is, *the cosine of the angle is unchanged by changing the sense* (or sign) *of the angle.* Now we learn in Algebra that only the even powers, not the odd powers, of a symbol are unchanged by changing the sign of the symbol; thus:

$$(-\alpha)^2 = \alpha^2, (-\alpha)^4 = \alpha^4, \text{ but } (-\alpha)^3 = -\alpha^3.$$

Hence the cosine is called an **even** function of the angle.

Its *periodicity* and its *evenness* are the two cardinal properties of the cosine, on which all others hinge.

321. We may now arbitrarily define the **sine** of an angle to be the **cosine of the complemental angle**; *i.e.* $90° - \alpha = \alpha\,|$, as we may write **sine of** α, which is commonly abbreviated into **sin** α.

---

* Two magnitudes such that the values of the one *correspond* to values of the other are called *functions* of each other.

Now write $90° - \beta$ for α;
we obtain
$$\cos\{90° - (90° - \beta)\} = (90° - \beta)|, \text{ or } \underline{\beta} = (90° - \beta)|;$$
i.e. *the cosine of an angle is the sine of the complemental angle.*

Hence the sine (or cosine) of either of two complemental angles, as α and β, is the cosine (or sine) of the other; *i.e.* if $\alpha + \beta = 90°$, then $\underline{\alpha} = \beta|$, and $\alpha| = \underline{\beta}$. When either changes by 2π, so does the other oppositely; hence the sine as well as the cosine returns into its original value; *i.e. the sine is also periodic with the period 2π.*

322. If the tract t be reversed, that is, turned through a straight angle, its projection will also be reversed, but otherwise unchanged; hence the cosine will be reversed; that is,
$$\cos(\alpha + \pi) = -\cos\alpha; \text{ also } \cos(\alpha - \pi) = \cos(\alpha + \pi) \text{ (why?)};$$
hence $\qquad \cos(\alpha - \pi) = -\cos\alpha,$

or, *to change the angle by the half-period, π, changes the sign of the cosine.*

323. Since the cosine is an even function,
$$\cos(\alpha - \pi) = \cos(\pi - \alpha);$$
hence $\qquad \cos(\pi - \alpha) = -\cos\alpha.$
That is, since α and $\pi - \alpha$ are supplemental, *the cosines of supplemental angles are counter* — equal in size, but opposite in sense (or sign).

324. Again, if $\alpha + \beta = 90°$, then $\alpha| = \beta$ (why?).
Then $\qquad (\alpha + \pi)| = \underline{\beta - \pi} \text{ (why?)} = -\underline{\beta} = -\alpha|.$
That is, *to change the angle by the half-period, π, changes the sense of the sine* as well as of the cosine.

244 GEOMETRY.

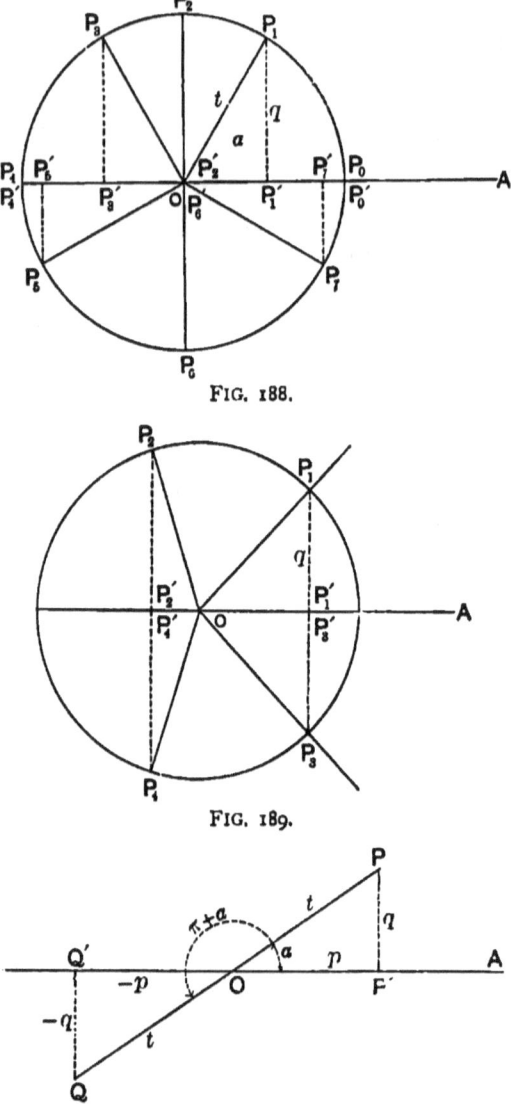

FIG. 188.

FIG. 189.

FIG. 190.

METRIC GEOMETRY. 245

We now ask, how is the sine affected by changing the sense of the angle? We have

$$(-\alpha)| = \underline{90° + \alpha} \text{ (why?)} = -\underline{(90° - \alpha)} \text{ (why?)} = -(\alpha)|;$$

that is, *the sense of the sine changes when the sense of the angle changes*. But this is the property only of *odd* powers, not of even powers, of a symbol;

thus $(-a)^3 = -a^3$, $(-a)^5 = -a^5$, etc.

Hence the sine is called **odd** function of the angle.

Its *periodicity* and its *oddness* are the cardinal properties of the sine, on which all others hinge.

325. If q be the projector of the tract t then $\dfrac{q}{t}$ is plainly the sine of the angle α for every position of t. We may indeed *define* the sine of α as equal to this ratio, and from this definition readily deduce all the foregoing properties (Figs. 188, 189, 190).

Exercises. 1. Prove that $(\alpha|)^2 + (\underline{\alpha})^2 = 1$.

2. Find the value of $\alpha|$ for $\alpha = 0, 30°, 45°, 60°, 90°, 120°, 150°, 180°, 210°, 225°, 240°, 270°, 300°, 330°, 360°, 390°$.

3. If a, b, c, be the sides of a $\triangle$, α, β, γ the opposite angles, r the circumradius, prove that

$$\frac{a}{\alpha|} = \frac{b}{\beta|} = \frac{c}{\gamma|} = 2r.$$

Such is the **Law of Sines** (Fig. 191).

4. Prove that $a^2 = b^2 + c^2 - 2bc\alpha$,—**Law of Cosines.**

5. If a and b are adjacent sides of a $\square$, and $\widehat{ab}$ denote the angle between them, prove that $\square = ab \cdot \widehat{ab}|$.

N.B. We may *define* the sine from this important theorem thus: *The sine of the angle between two sides of a $\square$ is that* number which taken as a multiplier turns the area of the rectangle of the sides into the area of the □.

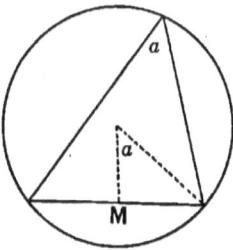

FIG. 191.

326. If we project successively the sides of a closed polygon on any ray, *the sum of the projections will be* 0; for the end of the projection of the last side falls on the beginning of the projection of the first. This fact is very important in surveying, where in compassing a field the sum of the northings must equal the sum of the southings, and these two sums, having opposite senses, together make the whole sum 0.

So for the eastings and westings.

We may express this fact in symbols thus :

$$s_1\underline{\alpha}_1 + s_2\underline{\alpha}_2 + s_3\underline{\alpha}_3 + \cdots + s_n\underline{\alpha}_n = 0 = \Sigma s\underline{\alpha}.$$

Here the s's are the sides, the α's are the angles of the sides with a fixed ray, as the east and west line, $s\underline{\alpha}$ is the projection of a side, and Σ is the symbol of summation.

Exercise. Show by projecting on a ray normal to the first ray that $\Sigma s\alpha | = 0$.

327. If we project consecutively all the sides of a polygon but one on that one, *the sum of the projections will be that one itself.* For the beginning of that side is the projection of the beginning of the first, and its end is the projection of the end of the last, of the projected sides (Fig. 192).

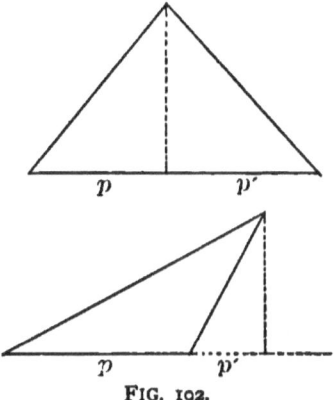

FIG. 192.

Thus, in a △, the sum of the projections of two sides on the third is the third.

328. We make a most important application of this simple fact in finding the *cosine of the difference of two angles* (Fig. 193).

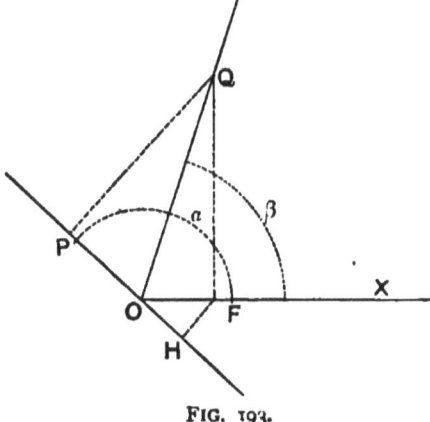

FIG. 193.

Let OP and OQ make angles α and β with any fixed ray OX. Then angle $QOP = \alpha - \beta$. Now project any tract OQ on OP; there $OP = OQ \cdot \underline{\alpha - \beta}$. But instead of projecting OQ directly we shall obtain the same result by projecting consecutively OF and FQ. The projection of OF is $OF \cdot \underline{\alpha}$, and of FQ is $FQ \cdot \alpha|$. But $OF = OQ \cdot \underline{\beta}$, and $FQ = OQ \cdot \beta|$.

Hence $OP = OQ \cdot \underline{\alpha - \beta} = OQ \cdot \underline{\alpha} \cdot \underline{\beta} + OQ \cdot \alpha| \cdot \beta|$.

Or, $\underline{\alpha - \beta} = \underline{\alpha} \cdot \underline{\beta} + \alpha| \cdot \beta|$.

By changing the sense of β and by putting $90 - \alpha$ instead of α, let the student show that $\underline{\alpha + \beta} = \underline{\alpha} \cdot \underline{\beta} - \alpha| \cdot \beta|$, $(\alpha + \beta)| = \alpha| \cdot \underline{\beta} + \underline{\alpha} \cdot \beta|$, $(\alpha - \beta)| = \alpha| \cdot \underline{\beta} - \underline{\alpha} \cdot \beta|$.

These four formulæ express the **Addition-Theorem** of Sine and Cosine.

The doctrine of Functions of Angles constitutes Trigonometry — an extremely important subject, which cannot be pursued any further here. See Smith's *Clew to Trigonometry*.

MEASUREMENT OF THE CIRCLE.

329. Thus far our linear measurements, or comparisons of length, have been wholly of tracts. The peculiar simplicity of such operations is due to the fact that any tract may be superposed (at least in thought) on any other, and thus their equality or inequality infallibly tested. We may similarly compare arcs of equal circles, but not arcs of unequal circles, nor arcs and tracts; for these cannot be made to fit on each other to even the smallest extent. We feel sure indeed that a circle or arc has a perfectly definite length, that it is longer than some tracts, shorter than others, and equal to some others. For if we suppose an inextensi-

MEASUREMENT OF THE CIRCLE. 249

ble cord wrapped around a circular disk, on unwinding and straightening the cord we should obtain a tract equal to the circle in length. But it remains difficult or impossible to fix the notion of the length or to determine the length itself without some preliminary definitions and assumptions.

We assume then that a circle has a definite length, neither more nor less; also, that it bounds a definite area, neither more nor less.

330. We now inscribe in the circle of radius r a regular n-side, and parallel to this latter we circumscribe a regular n-side; then bisecting each arc subtended by a side of the inscribed n-side we inscribe a regular $2n$-side and also circumscribe parallel to it a regular $2n$-side.

Then the following facts are at once evident:

1. The area of any inscribed polygon is less, and the area of every circumscribed polygon is greater, than the area of the circle; or if I_n, S, and C_n designate these areas, then

$$I_n < S < C_n \text{ (why?).}$$

2. The area of the inscribed $2n$-side is greater than that of the inscribed n-side, while the area of the circumscribed $2n$-side is less than that of the circumscribed n-side; or,

$$I_{2n} > I_n, \quad C_{2n} < C_n \text{ (why?).}$$

3. The area of each regular n-side is half that of the rectangle of the perimeter and the central normal on a side, and in case of the circumscribed polygons this normal is the radius r, but in case of the inscribed polygons this normal, or **apothem**, a_n, is less than r.

4. The perimeters of inscribed and circumscribed n-sides are to each other as a_n and r; for they are similar polygons, and a_n corresponds to r.

250   GEOMETRY.

5. Since the area of the circumscribed n-side decreases as n increases, and since one dimension, the radius, remains constant, it follows that the other dimension, the (half) perimeter, must decrease with increasing n. For $n = 4$ the polygon is a circumsquare, and the perimeter is $8r$.

6. Since the sum of two sides of a $\triangle$ is greater than the third side, it follows that the perimeter of the inscribed n-side increases with increasing n. For $n = 6$ the perimeter is $6r$. Hence for all higher values of n the perimeters of both inscribed and circumscribed polygons lie between $6r$ and $8r$.

7. Since then the sum of the n-sides is certainly less than $8r$, by making n large enough we can make each side, whether of inscribed or circumscribed polygon, as small as we please, smaller than one millionth, smaller than one billionth, smaller than any assigned magnitude however small.

8. But the half-side of the regular inscribed n-side is a geometric mean between the segments of the normal diameter; *i.e.*

$$r + a_n : \frac{S_n}{2} :: \frac{S_n}{2} : r - a_n.$$

As S_n is small at will, so is $\dfrac{S_n}{2}$, and still more is $r - a_n$, which we may call d_n, the distance between the parallel sides of the inscribed and circumscribed polygons.

N.B. A magnitude small at will is often called an **infinitesimal**. Since $r - a_n$ or d_n is infinitesimal with respect to $\dfrac{S_n}{2}$, which is itself infinitesimal, it (d_n) is called an infinitesimal of *2d order*. But we are not now concerned with this fact.

MEASUREMENT OF THE CIRCLE. 251

9. The difference in area between the circumscribed and the inscribed n-sides is a trapezoid the half-sum of whose parallel sides is less than $8r$, the altitude being d_n. Since $8r$ is finite and definite while d_n is small at will, it follows that the difference in area of the circumscribed and inscribed regular polygons is small at will.

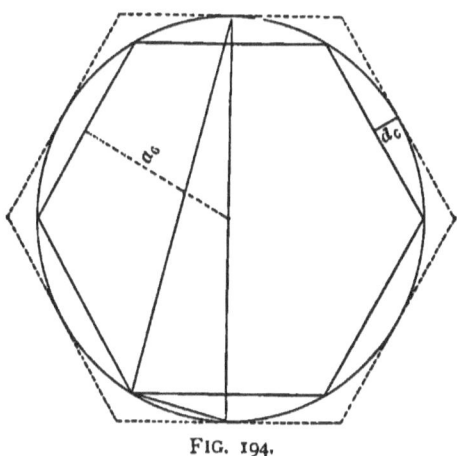

FIG. 194.

10. Again, we have from similar homothetic figures,

Perimeter of C_n : Perimeter of I_n : : $r : r - d_n$.

Hence, *dividendo*,

Perimeter C_n : Perimeter C_n — Perimeter I_n : : $r : d_n$.

But d_n is small at will; so too then is $D_n =$ Perimeter C_n — Perimeter of I_n; *i.e.* the difference in perimeter of the circumscribed and inscribed regular polygons is *small at will*.

11. The circle lying always wholly between the two polygons inscribed and circumscribed both in position and in areal value, it follows that the difference between its area

and the area of either polygon is small at will; and since the circle is fixed both in area and in position, it follows that the circle is the **Limit** of both polygons, and that the polygons close down upon it close at will as n grows ever larger and larger.

331. Think of the polygonal strip between the inscribed and circumscribed n-sides as growing ever narrower and narrower; the circle is a fixed *boundary* toward which the circumscribed n-side shrinks down as n increases, and in such a way that there is no assignable point outside of the circle, no matter how close to it, that will not also fall outside of the circumscribed n-side as n increases. Likewise the circle is the fixed *boundary* toward which the inscribed n-side swells out as the n increases, and in such fashion that there is no assignable point inside of the circle (no matter how close to it) that will not fall inside of the inscribed n-side as n increases. Thus the inscribed and circumscribed n-sides close down upon each other so as to leave no point between *except the points of the circle itself.* As n increases, the polygons *tend* to absolute coincidence with each other and with the circle. This fact is expressed fully and accurately by saying that the circle is the common **Limit** in length, in area, in position, of the inscribed and circumscribed regular n-sides for increasing n; it is expressed elliptically and inaccurately, but conveniently and frequently, by saying that the circle is or may be regarded as a *regular polygon of an infinite number of sides.*

332. The area of a circumscribed n-side is half the area of the rectangle of the radius of the circle and the perimeter of the polygon, for all values of n. Hence the area of the circle is half the rectangle of radius and the perimeter; that

MEASUREMENT OF ANGLES.

is, the circle. The numeric expressing the ratio of the circle to its diameter is called the *perimetric ratio*, and is designated by the Greek initial of perimeter, π. It is an irrational and hence not expressible exactly as a fraction, whether common or decimal, but its value has been calculated in various ages to various degrees of exactitude, and may be calculated to any degree of exactitude. Of late years it has been calculated (by Shanks) to the 707th decimal place, and verified to the 500th — a degree of accuracy immensely higher than can be attained in any measurement. For most practical purposes the value $\pi = 3.14159$ or even 3.1416 is close enough.

If then r be the radius, $2\pi r$ is the length of the circle, and $\pi r \cdot r$, or πr^2, is its area.

333. Because the ratio π is irrational it by no means follows that it cannot be constructed geometrically; that is, that we cannot with ruler and compasses draw a tract that shall be exactly equal to the circle of radius r. The ratio $\sqrt{2}$ is irrational, yet we can easily construct $\sqrt{2} \cdot r$, by drawing the diagonal of a square of side r. If we could draw a tract πr, equal to a half-circle, then, by Problem III, p. 193, we could construct the geometric mean of r and πr, which would be the side of a square precisely equal to the circle in area. This famous problem of *squaring the circle* is therefore not an irrational one; it is unsolved, but possibly not in itself unsolvable. But see Math. Ann., xx., p. 213.

MEASUREMENT OF ANGLES.

334. We have denoted by π the so-called *perimetric* ratio, namely, of the circle to its diameter, $2r$, so that, if r be the radius, then $2\pi r$ is the (length of the) circle, and πr^2 is its area. But there is another important use of π.

335. We have learned the ordinary sexagesimal division of the round angle into 360 equal parts called *degrees*. This *artificial* unit, degree, does not recommend itself for purposes of mathematical investigation, but a *natural* unit is suggested by the measurement of the circle itself. For it is plain that *whatever part an arc is of the whole circle, that same part the central angle of the arc is of the round angle.* Thus, if m times the arc make out the circle, then m times the central angle will make out the round angle; for, in adding the arcs about the centre O, we at the same time add the angles at O subtended by the arcs. If, however, arc and circle be incommensurable, cut the arc and also the angle into q very small equal parts. Then p of these arc-parts will be less and $p + 1$ will be greater than the circle, while, similarly, p of the angle-parts will be less and $p + 1$ will be greater than the round angle; and this will always hold, no matter how great q and p may be. Hence, using the arc as unit-arc and the angle as unit-angle, we see that the metric numbers of circle and round angle lie always between the same fractions, $\dfrac{p}{q}$ and $\dfrac{p+1}{q}$; and for increasing p and q these fractions *close down upon each other*, so that the metric numbers of circle and round angle cannot differ by ever so little, but must be precisely the same. If instead of circle and round angle we take any other arc and its corresponding central angle, the reasoning remains unchanged, so that we have

Theorem CXLVI. — *If any arc and its corresponding central angle be taken as units, the metric numbers of any other arc* (of the same or equal circle) *and its corresponding central angle are the same.*

In other words,

THE EUCLIDIAN DOCTRINE.   255

Central angles and their subtending arcs (in the same or equal circle) *are proportional* (see Art. 337).
The latter statement is conciser; the former is preciser.

336. Now a natural unit for arc-measurement is plainly *radius;* hence a natural unit for angle-measurement is *the angle whose arc is radius* (in length) ; accordingly we adopt it as unit-angle and name it **Radian**. The metric number of the circle, radius being unit, is 2π; hence the metric number of the round angle, radian being unit, is 2π. The radian equals about $57° 17' 33''$.

Corollary. If n be the metric number of any angle, and therefore of its corresponding arc, radian and radius being units, then of any other arc, subtending the same or equal angle, but described with a radius whose metric number is r, the metric number will be nr; for all circles are similar. That is,

The metric number of an arc equals the product of the metric numbers of its central angle and its radius.

Exercise. What are the natural metric numbers of a straight angle? A right angle? An angle of 60°? Of 45°? Of 30°? Of 120°? 150°? 225°? 240°? 270°? 420°? 600°? 720°? 1080°?

THE EUCLIDIAN DOCTRINE OF PROPORTION.

337. According to Euclid, *four magnitudes, a, b, c, d, are in proportion, taken in order, when any m-fold of the first is less than, equal to, or greater than, any n-fold of the second, according as the same m-fold of the third is less than, equal to, or greater than, the same n-fold of the fourth.*

In symbols

$$a:b::c:d \ (\text{read } a \text{ is to } b \text{ as } c \text{ is to } d)$$

when, and only when,

$$ma < nb, \ ma = nb, \text{ or } ma > nb,$$

according as

$$mc < nd, \ mc = nd, \text{ or } mc > nd.$$

We may also say equivalently that *a has the same ratio to b that c has to d when, and only when, etc.*

338. Hereby is defined, then, precisely, not indeed *ratio*, but at least *equality of ratios*. However, it still remains to be proved that the axiom of equal *magnitudes*, namely, *magnitudes equal to the same or equal magnitudes, are equal to each other*, can be applied to equal *ratios* ; for it has not yet been shown that ratios are magnitudes or may be treated as magnitudes. The all-important fact that, *whatever these ratios may be*, they obey the axiom of magnitudes, is expressed in the

Theorem CXI[a]. — If $a:b::c:d$ and $a:b::e:f,$

then $\qquad c:d::e:f$ (see Theorem CXI.).

For herein is declared that when two ratios, $c:d$ and $e:f,$ are equal to the same ratio, $a:b,$ they are equal to each other. *After* establishing this fundamental proposition, but *not before*, we may drop the double colon, : :, and write $a:b=c:d.$

339. We may illustrate both the idea and the method of Euclid, in demonstrating the following extremely useful

Theorem CXLVII. — *Areas of similar figures are to each other as the squares on homologous tracts* (in the figures).

TH. CXLVII.] THE EUCLIDIAN DOCTRINE.

Data: F and F' two similar figures, t and t' two homologous tracts, t^2 and t'^2 the squares upon them.

Proof. If the figures be curvilinear, and in general even if they be rectilinear, it will not be possible to cut them up into corresponding squares, however small, as is manifest. Nevertheless, we may cut each one up into congruent squares so small that the remainder shall be less than any assigned area, however small; that is, shall be *small at will*. For (Fig. 195) draw two corresponding series of equidistant horizontals and verticals, d apart in F, and d' apart in F'.

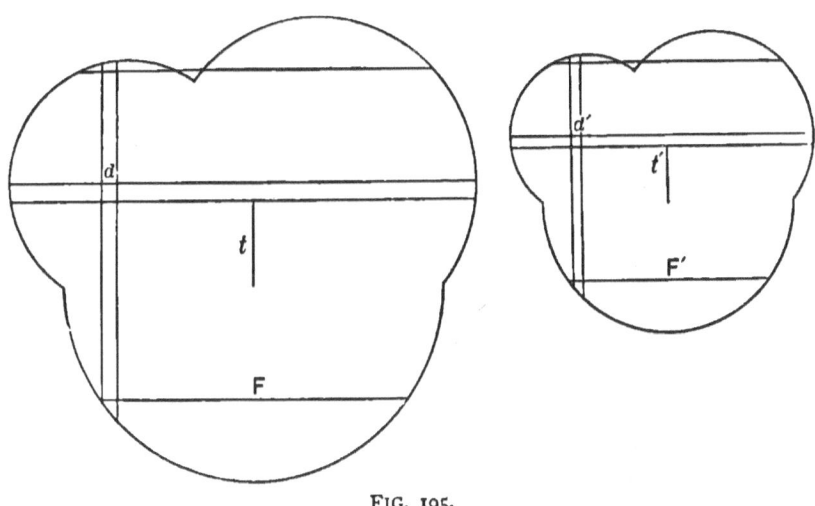

FIG. 195.

Let the extreme verticals in F be g apart, and let there be $v + 1$ of them, so that $g = v \cdot d$. Then the excess of the area of F over the sum of the squares will be the sum of the pairs of pieces at the end of each vertical strip; hence it will be less than $2\,v$ little squares, or less than a rectangle of base $2\,g$ and altitude d, *i.e.* $< 2\,gd$. But this rectangle is small at will, since its base $2\,g$ is constant and finite while its altitude d is small at will. Similarly, in F', $2\,g'd'$ is small as we please.

Now suppose there are p little squares cut out in F and also in F', where p is some very large but perfectly definite integer. Then the areas A and A' of F and F' will differ from the sums of the squares, pd^2 and pd'^2, by s and s', two magnitudes small at will; *i.e.*

$$A = pd^2 + s \quad \text{and} \quad A' = pd'^2 + s'.$$

Now suppose $md^2 = m'd'^2$; *i.e.* the sum of m squares in F equals the sum of m' squares in F'. Plainly then

$$p\,(md^2) = p\,(m'd'^2)\,; \quad \text{or} \quad m \cdot pd^2 = m' \cdot pd'^2 \quad (\text{why?}).$$

Now

$$mA = m \cdot pd^2 + ms \quad \text{and} \quad m'A' = m' \cdot pd'^2 + m's';$$
hence $\qquad mA - m'A' = ms - m's'.$

But s and s' are small as we please, while m and m' are finite; hence ms and $m's'$ are small at will; hence their difference, $ms - m's'$, is less than a magnitude small at will; hence *it is* 0. (Why? Because there is only one definite magnitude, namely, zero, that is less than a magnitude small at will.)

Hence, if $md^2 = m'd'^2$, then $mA = m'A'$.

Now suppose $md^2 > m'd'^2$. Then, as before, $mpd^2 > m'pd'^2$, and $mA = mpd^2 + ms$ while $m'A' = m'pd'^2 + m's'$.

Hence $\quad mA - m'A' = (m \cdot pd^2 - m' \cdot pd'^2) + ms - m's'.$

Here again $ms - m's'$ is small at will, while $m \cdot pd^2 - m' \cdot pd'^2$ is some finite magnitude; hence $mA - m'A'$ is also some finite magnitude of the same sense; *i.e.*

if $md^2 > m'd'^2$, then $mA > m'A'.$

Precisely in like fashion we prove that

if $md^2 < m'd'^2$, then $mA < m'A'.$

That is,
$$mA < m'A', \quad mA = m'A', \quad \text{or} \quad mA < m'A',$$
according as
$$md^2 < m'd'^2, \quad md^2 = m'd'^2, \quad md^2 > m'd'^2.$$

Hence, by definition, the areas are proportional to the squares on the corresponding tracts, d and d'; or
$$A : A' : : d^2 : d'^2.$$

Now compare the squares on t and t', d and d'. Since all squares are similar, and since d and d' are corresponding tracts or the equals of corresponding tracts in the squares on t and t', we have from the foregoing,
$$t^2 : t'^2 : : d^2 : d'^2.$$

Hence, by the Axiom of Magnitudes, applicable to ratios,
$$A : A' : : t^2 : t'^2. \qquad \text{Q. E. D.}$$

340. Care has been taken to conduct the foregoing demonstration so that it shall apply quite as well to circles, to regular triangles, in fact to any similar figures drawn on homologous tracts as to squares; so that we may affirm

Theorem CXLVII^a. — *Areas of similar figures are proportional to areas of any similar figures on homologous tracts* (in the original similar figures).

341. The peculiar propriety and advantage of using the **square** are seen on stating the analogous arithmetical

Theorem CXLVII^b. — *The metric numbers of similar figures are proportional to the metric numbers of any other similar figures in the same ratio of similitude.*

Now the figure (or area) whose metric number is easiest to find is the *square*, whose metric number is the *second power* of the metric number of its side. Instead of *second power* it is usual, almost universal, to say *square* and thus employ this latter term in two entirely different senses,— the proper geometrical sense and the tropical arithmetical sense. This double use of the term "square" is very regrettable as being especially confusing to beginners.

342. We may now restate our proposition thus:

The metric numbers of two similar figures are proportional to the metric numbers of squares on homologous tracts;

Or, are proportional to the second powers of the metric numbers of homologous tracts;

Or, are in the duplicate ratio of similitude of the figures themselves.

Thus, if the ratio of similitude be $2 : 5$ or $a : b$, the ratio of the areas will be $4 : 25$ or $a^2 : b^2$.

343. Observe carefully that the Euclidian doctrine of proportion is not a geometrical doctrine, but an arithmetical doctrine applied to Geometry. The same may be said of the accepted doctrine in modern texts: it is Arithmetic applied to Geometry. Nevertheless, the difference between the two is very great. Euclid's is based wholly on the operation of *multiplication* and employs only positive integers, not fractions nor irrationals, which indeed the Greek did not recognize as numbers; the modern, on the other hand, is based on the operation of *division*, and necessarily involves some general theory of fractions and irrationals. Moreover, Euclid's, while regarded as cumbrous and very difficult for the beginner, is yet a model of logical elegance and rigor: the like can hardly be said of the modern treatment.*

---

* The usual algebraical treatment of proportion is not really sound.
O. HENRICI, ENC. BRIT., VOL. X., *Geometry*, § 47.

The doctrine developed in this text is *purely geometrical*, implying no numerical knowledge or calculus. It is grounded in the notions of parallelism and similarity, to stand or fall with them. Hence it will be found to have no place in bi-dimensional spherics, the doctrine of the sphere-surface, which is in many particulars quite analogous to Planimetry, the doctrine of the plane, but in which there are no similar figures.

MAXIMA AND MINIMA.

344. Already, in Art. 135, the notions of maximum and minimum have been defined, but it is well to add here that not absolute but merely relative size is referred to, inasmuch as a varying magnitude may pass through a number of maxima and minima, and of these some maximum may be less than some minimum. Thus, a boy may inflate his

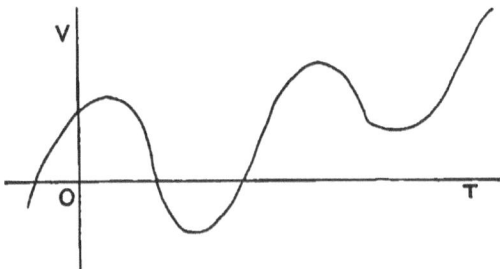

FIG. 196.

elastic balloon till the diameter becomes 9 inches, then let it shrink to a diameter of 7 inches, again inflate it to a diameter of 12 inches, let it shrink to one of 10 inches, again inflate it, and so on. Here 9 and 12 are maxima, while 7 and 10 are minima of the diameter. Plainly, maxima and minima alternate with each other, and the course of the

262   GEOMETRY.

variable may be depicted by a waving line, the values of the variable being the vertical distances of the points of the line from a fixed base-line. OV is the axis of the variable (Fig. 196); OT is the time-axis. What are the maxima and minima? How do the tangents lie at these points of the curve?

345. The general doctrine of maxima and minima calls for a method that shall seize upon the magnitude in the process of change, and subject its momentary variations to investigation. Such a method is supplied in the Infinitesimal Calculus. But there are many interesting and important geometric problems that yield even more readily and completely to elementary than to more refined methods, and some of these we shall now consider.

346. What is the maximum parallelogram with *given sides?* The student may easily show it to be a *rectangle*.

Corollary. The maximum triangle with two given sides is *right-angled* between those sides.

347. What is the maximum triangle with *given base* and *given vertical angle?* From the figure it is at once seen to

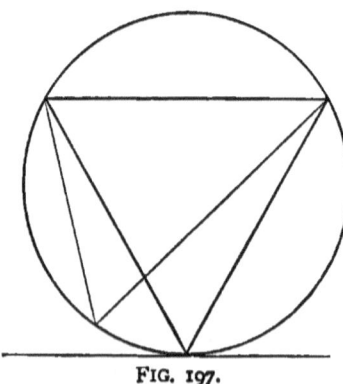

FIG. 197.

be *symmetric* (Fig. 197). Trace the variation in both area and perimeter of the △.

348. What is the maximum triangle with *given base* and *given perimeter?*

We are sure that the symmetric △ is either maximum or minimum; for as the vertex slips either rightward or leftward from the symmetric position by the same infinitesimal amount (Fig. 198), the two resulting ⚠ are congruent. Hence the

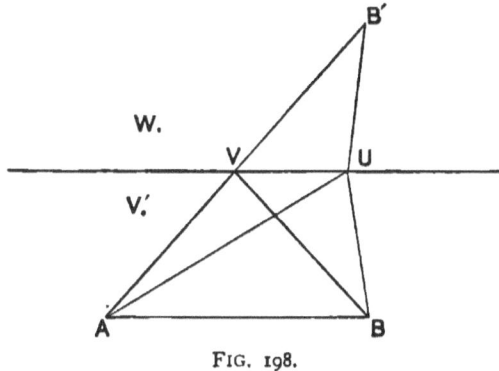

FIG. 198.

symmetric △, which lies between the two, is either greater or less than either. That it is greater, and hence is the maximum, is readily proved thus:

Through the vertex V draw a parallel to the base. Then no other position of the vertex can be on this parallel, as at U; for $AU + UB > AV + VB$, as is seen at once on taking the point B' symmetric with B as to the axis VU. Still less can the vertex take a position above the parallel, as at W. Hence it must in every other position be below the parallel, as at V'; and $AVB > AV'B$ (why?).

349. **Theorem A** (Lemma).— *If the base and perimeter of a rectilinear figure be given, the area may be continually increased by increasing the number of sides* (besides the base), *and keeping them mutually equal.*

Data: b the given base, s the sum of the other sides.

Proof. On b complete with s a symmetric △, a maximum. On either side take a point D distant one-third of the side AC from the vertex C. Now holding the new base BD fixed, convert BCD into a maximum (symmetric) △, keeping it *isoperimetric*, that is, of equal perimeter. The resulting quadrangle $ADEB$ is greater than the △ ACB (Fig. 199), and has its three sides AD, DE, EB equal. Now

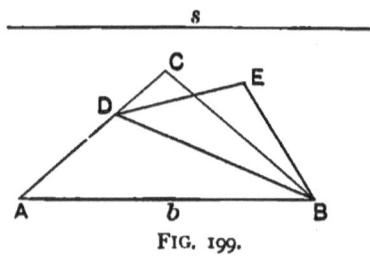

FIG. 199.

drawing AE, we may proceed similarly with the △ ADE; the resulting 5-side will be greater than the 4-side, but its sides will be unequal, and we can still further enlarge the figure, keeping it isoperimetric, by equalizing the four sides. Thus we may proceed continually, and at every step enlarge the bounded area, first increasing the number of sides, and then equalizing them. As long as any two consecutive sides are unequal, we can join their ends and enlarge their △ by making them equal. Q. F. D.

350. **Theorem B** (Lemma).— *Any n-side, the base being fixed and the other sides mutually equal, has maximum area only when the equal sides enclose equal angles.*

Data: We consider first a 4-side, AB its base, AC, CD, DB its three equal sides, and the angles C and D equal (Fig. 200).

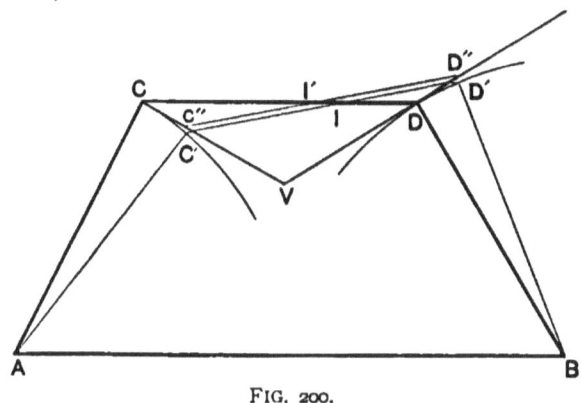

FIG. 200.

Proof. Suppose the sides to be rigid rods hinged at the angles, forming a *linkage*. Precisely, as in Art. 349, we know that the area is either a maximum or a minimum, and we easily prove it to be *the* maximum thus:

Deform the linkage by thrusting the hinge C down to C'; then *if* $AB > AC$, as C descends to C' on the arc of a circle, D will rise to D' on the arc of an equal circle.

1. Then *the arc* $CC' > arc\ DD'$. For $CD = C'D'$, and $C'D'$ being oblique, its horizontal projection is $< CD$; hence the lateral or horizontal thrust of C is $>$ the lateral thrust of D; that is, the horizontal projection of the arc or chord CC' is $>$ the horizontal projection of the arc or chord DD'. But even if DD' were only *equal* to CC', its horizontal projection would be *greater* than that of CC' (why?); hence DD' must be *less* than CC'.

Corollary. Hence $\triangle ACC' > BDD'$.

2. *The* $\triangle ICC' > IDD'$. For if the fixed length CD had slipped with its ends along the tangents at C and D into the

position $C''D''$, then we should have had $\triangle I'CC'' > I'DD''$ (why?); much more, when the ends slip along the arcs (or chords), the end C falls lower (to C'), the end D rises not so high (to D'), and the intersection slips further rightward (to I), the large subtractive $\triangle$ is increased to ICC', the small additive $\triangle$ is decreased to IDD'.

3. Hence the two decrements ACC' and ICC' produced by this deformation being respectively greater than the two increments BDD' and IDD', it follows that the resultant area $AC'D'B$ is less than the original area $ACDB$. The reasoning is not changed if we thrust in D instead of C, and it applies whatever the amount of the thrust. Hence the anti-parallelogram $ACDB$ is the maximum.

If, however, $AB < AC$, then the argument about the arcs CC' and DD' is no longer valid (why?). But in this case it suffices to use CD instead of AB as the fixed base, and then proceed precisely as before.

If $AB = AC$, then the anti-parallelogram becomes a square, and is plainly larger than the rhombus resulting from any deformation (why?).

Hence in all cases the symmetric 4-side is greater in area than any asymmetric one.

If, now, in any n-side with $(n-1)$ equal sides, any two consecutive angles between equal sides be unequal, as PQR and QRS, then we may apply the preceding reasoning to the 4-side $PQRS$, and increase its area by making it an anti-parallelogram; and so we may proceed continually, enlarging the area as long as the angles between the equal sides are not all equal. Moreover, if these angles be all equal, then any change in the size of any one must change the size of one adjacent, and hence decrease the area of a 4-side, and therewith of the whole n-side. Hence the symmetric n-side, with $(n-1)$ equal sides, $(n-2)$ equal angles, and two other equal angles, is *the* maximum. Q. E. D.

351. Such an n-side, however, is *always encyclic* (why?), and we have seen that its area may be continually increased by increasing n; hence for no finite value of n can the area be an absolute maximum. As n increases, the perimeter tends accordingly towards a circular arc as its *limit*, and the area always increasing tends towards the absolute maximum as its *limit*. Moreover, the polygonal area may be made to differ from the circular by *less* than σ *small at will*, while any change from the circular shape will produce some definite decrease in area at least equal to σ (for any such change may be brought about by a definite change however small in the polygonal area followed by other changes small at will). Hence in the *circular form the area attains its absolute maximum*.

352. The foregoing reasoning preserves its cogency however small the base b may be, and even when it vanishes, as is manifest, so that we have as *special cases*:

1. Of all isoperimetric n-sides the regular has maximum area.

2. A regular n-side has greater area than the isoperimetric regular $(n-1)$-side.

3. Of all isoperimetric figures the circle has maximum area.

353. The conclusions reached with respect to maximal areas of isoperimetric figures may now be readily converted into another set of conclusions with respect to minimal perimeters of equiareal figures; thus:

1. Of all equiareal n-sides the regular has minimum perimeter.

2. A regular n-side has less perimeter than the equiareal regular $(n-1)$-side.

3. Of all equiareal figures the circle has minimal perimeter.

The details are left to the student, who must remember that under fixed conditions increase of perimeter brings increase of area (why?).

The doctrine of Maxima and Minima is one of the most beautiful and fascinating in the whole range of mathematics, and especially in its applications in Mechanics it is of the highest practical as well as theoretical interest.

CONCLUDING NOTE.

354. Before closing our discussion it may be well to recall attention to certain matters of vital significance for geometric theory, but which could not be adequately treated earlier without perplexing and even revolting the student. The following sections make no pretension to thoroughness, but may yet enable the reader to orient himself properly in the subject.

355. Some diversity of judgment prevails as to what is the simplest form that can be given to the fundamental assumptions of Geometry. Euclid's so-called Axioms, comprised in his ' common notions,' Postulates, and Definitions, assume :

Continuity and the possibility of Rigid Motion (that is, of moving a body or figure without changing its size or shape) ;
The Existence of Surfaces, Lines, and Points ;
The Existence of Planes, Straight Lines, and Circles ;
The Existence of one, and only one, Non-intersector of a straight line for every point in its plane ;
The Equality of all Right Angles.

This last may be proved, and hence is unnecessary. To the others must be added :

The Infinity of the Straight Line,—not mentioned by Euclid, but yet implied in his demonstrations, and inserted by his editors.

Perhaps most moderns would prefer to assume openly the Existence of Plane, Straight Line, and Circle. The deduction of these notions from that of the sphere, given in this text in the main after Bolyai and Frischauf, even though it may "lay no claim to absolute rigor," nevertheless seems to the writer to be the most natural and easy to intuit.

356. The term **ray** has been preferred to *straight line*, or 'straight' (Halsted), because it seems important to have a single word for such a fundamental concept, and still more because the adjective 'straight' involves an unfortunate and unnecessary assumption. Rays may not be absolutely straight, but may curve with space itself, and return upon themselves like the Equator. They are not necessarily *straight*, but *as straight* as can be, the *straightest* that can be, in our actual space. Think of one end of a short string fastened on an egg-shell, and the string stretched over the shell by a weight at the other end. The string would then mark a *straightest* (geodetic) line *on the shell*, which would not, however, be straight.

357. The infinity of the ray, though not openly affirmed in the text, is yet implicit in certain demonstrations. Thus, in Theorems XVII. and XXVIII., it is assumed that we can lay off on the medial ray a tract equal to the medial tract without returning through the vertex. But we cannot certainly do this unless the ray be infinite. A meridian, familiar from Geography, is the straightest line that we can draw on a sphere, and corresponds to the ray

in a plane; but if we go more than half a meridian from the North pole, and then go on as far-again on the same meridian, we shall always return through the same North pole. Hence the demonstrations in the text fail in generality, if the ray be of finite length. Hence, too, the familiar and plausible Theorem XXIII., B, that from a point without a ray *only one* normal can be drawn to the ray, also fails unless the ray be infinite; for the proof of it rests on Theorem XVII. In fact, on a sphere all meridians from a certain point, the pole (North or South), are normal to the Equator, the *straightest* line on the sphere. In like manner the reasoning of Arts. *65 and *66, after Bolyai and Frischauf, is seen to assume the infinity of the ray and the plane,— let the student show at what points.

358. *But in the more general discussion* of Arts. 67–71 all such assumptions — as well as Axiom 7, that two rays can meet in only one point, which is *known* to hold only for the comparatively small region of our experience — are dropped, and the student must note with the utmost care that **four** possible space-forms result from our refusal to make these assumptions:

A. If the rays LM and $L'M'$, isoclinal to the third ray or transversal T, meet in two points, on the right and on the left, then we have so-called **double Elliptic** space.

B. If they meet in one point only, then we have **simple Elliptic** space.

C. If they do not meet at all, but if there be also other non-intersectors through the point O' (that is, if we grant Axiom A but reject Axiom B), then we have **Hyperbolic** space.

D. Lastly, if we grant both Axiom A and Axiom B, then we have ordinary **Parabolic** space.

359. The names Elliptic, Hyperbolic, Parabolic have been given by *Klein*. They mean *lacking, exceeding, equalling*, and refer to a certain characteristic magnitude called the Riemannian 'measure of curvature' (Riemann'sche *Kruemmungsmaass*), which in the three cases is respectively negative, positive, 0, or less than 0, greater than 0, equal to 0. Instead of Klein's terms we sometimes meet with **Riemannian, Lobatschevskian** (or *Gaussian*), and **Euclidian**, from Riemann, Lobatschevsky and Gauss, and Euclid, — mathematicians that first set forth clearly the properties of the space-forms.

360. Some of the distinguishing features of these four spaces are the following :

A. 1. The ray is closed and finite.
2. The sum of the angles in a plane $\triangle$ is $>$ a straight angle.
3. Two rays that meet in one point meet also in a second point.

B. 1. The ray is closed and finite.
2. The sum of the angles in a plane $\triangle$ is $>$ a straight angle.
3. Two rays meet at most in one point only.

C. 1. The ray is not closed, but infinite.
2. The sum of the angles in a plane $\triangle$ is $<$ a straight angle.
3. Two rays meet at most in one point only.

D. 1. The ray is not closed, but infinite.
2. The sum of the angles in a plane $\triangle$ is $=$ a straight angle.
3. Two rays meet at most in one point.

272   GEOMETRY.

361. It is curious and noteworthy that the ray in a *simple* Riemannian plane cuts the plane through, but *not in two*. For, take any ray R and two points close together, P on the left and Q on the right of R; through P and Q draw a ray meeting R at M. Then we may pass from P to Q rightward through M; or, since the rays are closed and meet only in M, we may pass from P to Q leftward and not through M; *i.e.* we may pass from one side of the ray R to the other without crossing it. This may be hard for us to imagine, but perhaps not harder than for the ancients to imagine antipodes. Think of a hollow ring — a circle running all round it or across it would cut it through, but not in two. The Riemannian plane is not such a ring, but is ring-like in being thus *doubly compendent* (Art. 162).

362. It will be well for the student to observe carefully just where the proof of Theorem XXXI. breaks down on rejecting Axiom B. We may then still draw through C a non-intersector of AB, making the alternates a and a' equal; and we may also draw through C a non-intersector of AB, making the alternates β and β' equal. But in the *absence of Axiom B*, we cannot know that these two non-intersectors are the *same;* hence we cannot know that the sum $a' + \gamma + \beta' =$ a straight angle; hence we cannot affirm that the sum $a + \gamma + \beta =$ a straight angle. In fact, if the two non-intersectors are not the same, then manifestly $a + \gamma + \beta = a' + \gamma + \beta' <$ a straight angle.

363. Actual measurement, direct and indirect, of the angles of a $\triangle$ yields a sum always *very* near to a straight angle. But even the largest $\triangle$ we can construct and compute in the heavens are yet *extremely* small relatively to the whole of space, whether space be finite or infinite; and since the defect under a straight angle, or the excess over a straight angle, if there be any such defect or excess, must vary with the size of the △, in observed △ it would be extremely small and so might elude our observation. In fact, for extremely small △, the only △ of experience, the four spaces are so nearly alike in properties as to be indistinguishable ; just as if the earth's radius were a decillion times as great as it is, and our experience extended over no more than its present surface, we should be unable to say whether it was flat, or sphere-shaped, or egg-shaped, or ring-shaped, or saddle-shaped.

364. It appears then that the natural question, Which of the four possible homœoidal spaces is our actual space ? is at present unanswerable. Our experience is still too narrow to enable us to decide or even to conjecture. Why, then, do we seem to prefer parabolic space, and build up our geometries on Euclid's foundations ? *Because it is easier, more convenient.* The superior simplicity of the Euclidian geometry is conspicuous in its doctrine of the parallel, the *unique* intersector, and of the sum of the angles in the plane △, which is a *constant*, the straight angle. The ground of our preference, then, is not a logical, but an economical one.

365. Lastly, let the student never forget that the question as to the fundamental properties of our space is at bottom a question as to the constitution of our own minds. It is they that at every instant project images of their own states and of all beings as related to them, and build up these projections into the world of phenomena about us, which we call space and its contents. Space, then, is made the way our spirits make it, and to know its fundamental properties is to know fundamentally the mode in which the spirit objectifies to itself, makes an object of its own contemplation, the world of Not-self about it. Whence it appears that

not only all physical problems, but also all geometrical problems, root finally in Metaphysics.

366. In the writer's judgment the doctrine of non-euclidian spaces and of hyper-spaces in general possesses the highest intellectual interest, and it requires a far-sighted man to foretell that it can never have any practical importance.

The student who would pursue the subject should read Halsted's excellent translations of Lobatschevsky and Bolyai, the Lectures and Addresses of Clifford and Helmholtz, Ball's article on *Measurement* in the Encyclopædia Britannica, and afterwards the monographs of Riemann, Klein, Newcomb, Beltrami, Killing, and should also consult the bibliography of the subject as given by Halsted in the *American Journal of Mathematics*, Vols. I. and II.

EXERCISES V.

1. Central rays of a parallelogram bisect it, and conversely.
2. Central rays of any closed centrally symmetric figure bisect it, and conversely.
3. Tracts bisecting the mid-parallel of a trapezoid and ending in its parallel sides bisect it, and conversely.
4. The sums of the opposite ⚠, into which tracts from a point within a parallelogram to its vertices divide it, are equal.
5. The sums of the opposite ⚠ formed by tracts from any point of the mid-parallel to the vertices of a trapezoid are equal.
6. Tracts from a vertex of a regular 6-side to the other vertices and the mid-points of the remotest sides divide it into six equal ⚠.
7. The △ whose vertices are the ends of one oblique

side of a trapezoid and the mid-point of the other is half the trapezoid.

8. The $\triangle$ of the medials of a $\triangle$ has $\frac{3}{4}$ of the area of the $\triangle$.

9. P is a point, $ABCD$ a parallelogram; then $\triangle APC = \triangle APD \pm \triangle APB$.

10. The joins of the mid-points of the opposite sides of a 4-side concur with the join of the mid-points of the diagonals.

11. *Divide a $\triangle$ or parallelogram into two parts in the ratio $l:m$.

12. Divide a parallelogram by tracts from a vertex into n equal parts.

13. Divide a $\triangle$ into n equal parts by tracts from a point on a side.

14. Transform a $\triangle$ or parallelogram into another of same base with a given angle at the base.

15. Transform a $\triangle$ or parallelogram into another with given base.

16. Transform a $\triangle$ or parallelogram into another with given base and given adjacent angle.

17. Transform a trapezoid into an anti-parallelogram of the same altitude.

18. The common border of two areas lying between two rays is a broken line; rectify this border without affecting the areas.

19. Inscribe in a given circle a rectangle of given area.

20. If tracts between the parallel sides of a trapezoid, not intersecting within it, divide the mid-parallel into n equal parts, they also divide the area into n equal parts.

21. In a trapezoid, the join of the mid-points of the diagonals equals the difference of the parallel sides.

22. Construct two tracts, knowing their ratio and their sum or difference.

23. Investigate the proportionalities between the seg-

ments of the altitudes of a △ made by the orthocentre and the segments of the sides made by the altitudes.

24. To the base of a △ draw a parallel cutting off a △ of given perimeter.

25. *The sum of two similar figures on the sides of a right △ equals a third similar figure on the hypotenuse.

26. If intersecting semicircles be drawn on all the sides of a right △, the sum of the two small crescents will equal the larger (Hippocrates, 450 B. C.).

27. Bisect a △ by a parallel to its base, and generalize the problem.

28. The rectangle of the segments into which one altitude of a △ is cut by the others is the same for all the altitudes.

29. Construct a △, knowing its altitudes.

30. The medials of a △ cut it into six equal △.

31. △'s mid-rays cut sides into $ir js lt$; $\Sigma ij = ij + jl + li = \Sigma rs$.

32. The altitudes of a △ cut it into six △, so that the sums of the alternate △ are equal.

33. Through the perimeters of an inscribed and a circumscribed regular n-side express those of the $2\ n$-side.

34. The area of a ring between two concentric circles equals that of a circle having as diameter the chord of the outer circle tangent to the inner circle.

35. The radius of the earth being 6370 kilometers, how far might one descry a ship from the summit of Chimborazo (21,424 feet)?

36. How many hills of corn one yard apart may be planted in a rectilinear field of one acre?

37. Chords from any point of a circle to the ends of a diameter divide its conjugate chords harmonically.

38. Chords from any point of a circle to the ends of a chord divide its conjugate diameter harmonically.

39. Transform a △ into another △' similar to a given △".

40. Find the locus of a point whose distances from two given rays are in a fixed ratio.

41. All rays whose distances from two fixed points are in a fixed ratio envelop two fixed points; find them.

42. On a given ray find a point whose distances from two fixed points (or rays) are in a fixed ratio.

43. Through a given point draw a ray whose distances from two fixed points shall be in a fixed ratio.

44. Find a point (or points) whose distances from three fixed non-concurrent rays L, M, N shall be as $l:m:n$.

45. Find a ray (or rays) whose distances from three non-collinear points A, B, C shall be as $a:b:c$.

46. Find a point whose distances from three fixed points shall be as $l:m:n$.

47. In every trapezoid the mid-points of the bases along with the intersections of the non-parallel sides and of the diagonals form an harmonic range.

48. The locus of a point whose distances from A and B are in the ratio $a:b$ is a circle on AB as central ray and dividing AB harmonically in the ratio $a:b$.

49. The locus of a point whence the collinear tracts AB and BC appear equal is a circle dividing AC harmonically.

50. Find the point whence three collinear tracts AB, BC, and CD appear equal.

51. P is a given point in a given angle BAC; draw BC so that $PB:BC::l:m$, or $BP:PC::l:m$, or $BA:AC::l:m$.

52. From the continued ratio of the sides determine the continued ratio of the altitudes of a △.

53. A marble rests in a conical glass; another rests on it and touches the glass all around; and another, and so on. How are the radii of the marbles related in size?

54. Find the area bounded by three equal tangent circles.

55. Compare the $\triangle$ of the centres with the $\triangle$ of the common tangents of three tangent circles.

56. In a regular $\triangle$ is inscribed a circle; tangent to it and two sides of the $\triangle$, another circle; and so on. How do the radii decrease?

57. M is the mid-point and P any other point of the tract AB; semicircles are drawn on $AM, MP, AP,$ and PB, all on the same side of AB; show that the sum of the first and second equals the fourth plus the area bounded by the first three.

58. Inscribe a circle in a given sector of a circle.

59. Find a point on a circle whose distances from two chords are proportional to the chords.

60. The distance of any point of a circle from the chord of contact of two tangents is a mean proportional between its distances from the tangents.

61. When does the altitude to the hypotenuse of a right $\triangle$ divide the hypotenuse in extreme and mean ratio?

62. In a regular 5-side each diagonal is divided in extreme and mean ratio by two others and all form a regular 5-side. Compare the areas of the two 5-sides.

63. The rectangle of the distances of any point of a circle from two opposite sides of an encyclic 4-side equals the rectangle of the distances from the diagonals.

64. RA and RB are two equal rigid rods pivoted at R; OQ is a rigid rod pivoted at O and $OQ = OR$; $AQBP$ is a rhombus formed of equal rigid rods movable about their junction-points. Show that, as Q traces a circle about O, P traces a ray normal to the ray OR, and that P and Q are inverse as to R.

N.B. Such is the **mechanical invertor** called **Peaucellier's cell.**

EXERCISES V. 279

65. In any range of four points, as $ABCP$, show that $AB \cdot CP + BC \cdot AP + CA \cdot BP = 0$.

66. In any pencil of four tracts, as OA, OB, OC, OD, show that
$\triangle AOB \cdot \triangle COD + \triangle BOC \cdot \triangle AOD + \triangle COA \cdot \triangle BOD = 0$.

67. The diagonals of a parallelogram concur with the diagonals of its complemental parallelograms.

68. The mid-points of the diagonals of a 4-side are collinear.

69. Rays through the vertices of a $\triangle$ and the points of touch of the ex-circles concur.

70. Normals to the sides of a $\triangle$, at the points of touch of the ex-circles, concur.

71. When normals from the vertices of one $\triangle$ to the sides of another $\triangle'$ concur, so do normals from the vertices of $\triangle'$ to the sides of $\triangle$.

72. Normals to the three sides of a $\triangle$ through the points of touch of two ex-circles and the in-circle concur.

73. The feet of normals from any point of a circle to the sides of an inscribed $\triangle$ are collinear (on **Simson's Line**).

74. Rays through the vertices of a $\triangle$ and the points of touch of the in-circle concur.

75. Mid-rays of the angles of a $\triangle$ (three outer, or two inner and one outer) intersect the opposite sides collinearly.

76. Tangents to the circumcircle of a $\triangle$ at its vertices intersect the opposite sides collinearly.

77. If the joins of the vertices of a $\triangle$ with three intersections of the opposite sides by a circle concur, so do the joins of the vertices with the other three intersections of the opposite sides.

78. Find a circle as to which two pairs of collinear points are inverse. (*Hint.* Draw a circle through each pair, and also their power-axis.) When is the circle sought *real*?

79. Find the inverse of a given point as to a given circle.

80. Given two points, find a circle of inversion having given radius or given centre.

81. A circle through a pair of inverse points cuts the circle of inversion orthogonally, and conversely.

82. The squared distances of a point on the circle of inversion from two inverse points are in the ratio of the central distances of the points.

83. The power-axis of a fixed circle and a variable circle through two inverse points as to the fixed circle envelops a fixed point.

84. Find the inverse of a circle when it passes through and also when it does not pass through the centre of inversion.

85. When does a circle invert into itself?

86. A circle, its inverse, and the circle of inversion have a common power-axis.

87. Inversion does not change the size of the angle under which two lines intersect.

88. The 9-point circle of a △ touches the in- and ex-circles of the △.

89. The joins of the intersections of two circles with their common diameter and a common orthogonal circle concur on that orthogonal.

90. The join of the polars of two points is the pole of the join of the points.

91. If a pole trace the sides of an n-side, P, the polar will envelop the vertices of an n-angle, Q; and if the pole trace the sides of Q, the polar will envelop the vertices of P.

92. If the pole trace any line L, the polar will envelop a corresponding line L'; and if the pole trace L', the polar will envelop L.

EXERCISES V.

Defs. Two lines, either of which is enveloped by the polar, as the other is traced by the pole, are called **reciprocal**.

When the reciprocals coincide, the figure is called **self-reciprocal** or **self-conjugate**, while the centre and the circle of reference are called the **polar centre** and the **polar circle** of the figure.

93. If a △ has a polar centre, it is the orthocentre.

94. If the polar circle is real, the △ is obtuse-angled.

95. The polar circle of a △ is orthogonal to the circles on the sides as diameters.

96. Invert the sides of a △ as to its polar circle.

97. The ends of a diameter of a circle are conjugate as to every orthogonal circle.

98. As the orthogonal circle (of 97) varies, the polar of either diameter-end envelops the other.

99. The distances of any two points from a polar centre vary as the distances of each from the polar of the other as to that centre (Salmon).

100. Polar reciprocal △ are in perspective.

Def. Circles, every pair of which have the same power-axis, are called **co-axal**.

101. If two circles intersect in A and B, all co-axals go through A and B.

102. The contrapositive of 101.

Hence there are two kinds of co-axals: *common point* co-axals (101) and non-intersectors, so-called *limiting point* co-axals (102).

103. The centre line of common point co-axals is a ray, namely, the common power-axis of co-axals orthogonal to the common point co-axals.

104. One and only one of these orthogonals goes through every point of the plane *except* the common points, which are therefore called (Poncelet) **limiting points** of the orthogonal co-axals.

105. Draw a double system of mutually orthogonal co-axals.

106. The polars of a point as to co-axals concur, and conversely.

107. The difference of the powers of a point as to two circles is proportional to the distance of the point from the power-axis.

108. The locus of a point whose tangent lengths from two circles are in fixed ratio is a co-axal circle.

109. When does the power-centre of three circles become indefinite, and when does the radius become imaginary?

110. The polar centre of a $\triangle$ is the power-centre of three circles on any tracts from the three vertices to the sides as diameters.

111. Three collinear points on the sides of a $\triangle$ being joined with the vertices, the circles on these joins as diameters are co-axal.

112. The four polar centres of the four $\triangle$, formed by the sides of a 4-side taken in threes, are collinear on the power-axis of circles on the diagonals of the 4-side as diameters.

113. Invert a double orthogonal system of co-axals; as special case take a common or limiting point as centre.

114. Show that any two circles may be inverted into equal circles, and find the locus of the centre of inversion.

115. Invert three non-co-axal circles into three equal circles.

116. Draw a circle touching these three equal circles, and re-invert the four circles. What problem is hereby solved?

117. Discuss the various possible positions of the centres of similitude of two circles.

Def. The **circle of similitude** of two circles has the tract between the centres of similitude as diameter.

118. The circle of similitude is co-axal with the two circles.

119. From any point on the circle of similitude the two circles appear to be of the same size.

120. Find a circle as to which the power-axis and the circle of similitude invert into each other.

Defs. A tract with its ends on a figure is called a **chord** of the figure. A ray bisecting a system of parallel chords is called a **diameter** of the figure. Two diameters, each halving all chords parallel to the other, are called **conjugate**.

121. Every two-ray (*Zweistrahl*), or angle considered not as a magnitude but as a figure, has an infinity of diameters, namely, every ray through its vertex.

122. A △ has three diameters, namely, its medials.

123. A parallelogram has two pairs of conjugate diameters.

124. A two-ray has an infinity of conjugate diameters.

125. If L' and M' be conjugate diameters of (the two-ray) LM, then L and M are conjugate diameters of (the two-ray) $L'M'$.

Def. Four such rays are called **harmonic**, because:

126. They cut every transversal in four harmonic points.

127. Conversely, four concurrent rays through four collinear harmonic points are harmonic.

128. The join of an outer vertex with the intersection of the inner diagonals of a 4-side cuts two opposite sides, each in a fourth harmonic to the three vertices on the side.

Hint. In Fig. 54, let I be the intersection of the inner diagonals CE, DF; let a 4th harmonic through A cut BC and BD at H and K. Then ID, IE, IB, IK are four harmonic rays, and so are IF, IC, IB, IH; also ID, IE, IB, are the same rays as IF, IC, IB; hence IH and IK are the same ray; hence the 4th harmonic through A goes through I.

129. Enumerate the harmonic ranges and pencils in 128.

130. A system of co-axals determines the points of every ray in pairs of conjugates, P and P', Q and Q', so that

$IP \cdot IP' = IQ \cdot IQ'$, where I is the intersection of the ray with the power-axis.

Defs. Points so determined are said to form an **Involution**. The fixed point I is called the **centre** of the Involution, the constant product or rectangle of the central distance of the conjugates is called the **power** of the Involution, and is positive or negative according as the distances are like-sensed or unlike-sensed.

131. An Involution is determined by its centre and a pair of conjugates, or by two pairs of conjugates.

132. When the power is positive there are two self-conjugate points (called *double points* or **foci**), and the focal tract is divided harmonically by every pair of conjugates.

133. *Conversely*, all pairs of points dividing a tract FF' harmonically form an Involution with F and F' as foci and the mid-point I of FF' as centre.

134. When the power is negative there are no (real) foci but there are two conjugate points, E and E', equidistant from the centre.

Def. Rays of a pencil passing through an Involution of points form an **Involution of Rays**.

135. Develop and express the reciprocal properties corresponding to 131–4.

136. *In general*, the points of a row and the intersections of their polars with the axis of the row form an Involution.

Hint. P the point, L the axis, O the centre of the referee circle S, which cuts L at F and F', Q the intersection of L with the polar of P, $OI = d =$ distance of L from O, $OA = r =$ radius of S on OP. Draw a circle K on PQ as diameter about C cutting OP at R. Then, difference of powers of O and I as to K is
$$\overline{OC}^2 - \overline{IC}^2 = \overline{OI}^2 = r^2 - IQ \cdot IP.$$
Hence $r^2 - d^2 = IQ \cdot IP =$ a constant. Q. E. D.

137. What exception does 136 suffer? What are the relations of the three possible cases?

138. Two polar conjugate points (or rays) and the pole (or polar) of their join determine a polar reciprocal $\triangle$.

139. Conjugate points on a ray (or rays through a point) form an Involution. When positive? When negative?

140. Two conjugate points in a secant divide the chord, and two conjugate rays through a point divide the angle of tangents from the point, harmonically.

141. The outer vertices and the intersection of inner diagonals of an encyclic 4-side form a polar $\triangle$.

142. The diagonals of a pericyclic 4-side form a polar $\triangle$.

143. Employ 141 and 142 to find the polar of a given pole and the pole of a given polar by use of ruler alone.

144. Given a centre of similitude of two circles, find its polars as to the circles and the power-axis by use of ruler alone.

145. Given the power-axis of two circles, find its poles as to the circles and the centres of similitude by use of ruler alone.

Defs. Two figures are said to be **in perspective** where the joins of corresponding points all go through a point called the **centre of projection**. — The rays are called **rays of projection**. Parallel rays are thought concurrent at ∞. — Two pencils are said to be **in perspective** when the joins of corresponding rays all lie on a ray, called the **axis of projection**. — The centre of projection and the axis of projection are plainly **self-correspondent**. — Each system of points is also said to be *in perspective* with the pencil of rays through them. — In elementary work we impose the condition that collinear points in the one figure shall correspond to collinear points in the other.

146. Two tracts are always in perspective as to two centres.

147. Express and prove the reciprocal theorem as to angles (two-rays).

148. Three collinear points and three concurrent rays are always **projective**, *i.e.*, may always be *brought* into perspective.

149. Two triplets of collinear points are always in projection. (For it is enough to slip the triplets each along its ray till a pair of correspondents coincide.)

150. State and prove the reciprocal theorem for pencils of three.

Def. The ratio of the distances of any third ray N of a pencil from two base-rays L and M of the pencil is called the **distance-ratio** of the third ray as to the other two; it may be written $\left(\dfrac{LN}{NM}\right)$ and is reckoned $+$ or $-$ according as the angles LN and NM are reckoned in the same or in opposite wise.

151. Trace the course of the distance-ratio as the third ray completes a rotation about the centre of the pencil.

152. The distance-ratio equals the **sine-ratio** of the angles formed by the third ray with the base-rays.

153. When two distance- or sine-ratios are counter, the four rays are harmonic, and the ratio of their ratios is -1.

154. Compare distance-ratios of corresponding angles and tracts in a pencil.

Def. The ratio of two distance-ratios in the same pencil, whether of tracts or of angles, is called the **cross-ratio**, or ratio of double section (*Doppelschnittsverhaeltniss*), or anharmonic ratio, of the bounding points or rays, and is written $(ABCD)$ or $(LMNP)$. Its value is

$$\frac{AB \cdot CD}{BC \cdot DA} \quad \text{or} \quad \frac{\sin \widehat{LM} \cdot \sin \widehat{NP}}{\sin \widehat{MN} \cdot \sin \widehat{PL}}.$$

EXERCISES V. 287

155. The joins of the vertices of a △ with three collinear points on the opposite sides divide the angles so that the triple sine-ratio $= -1$.

156. Three concurrent rays through the vertices of a △ divide the opposite sides so that the triple distance-ratio $= +1$ (*Ceva*, 1678).

157. Convert these two theorems and those of Arts. 315, 316.

158. Apply these converses in establishing the concurrences of altitudes, mid-normals, mid-rays, medials, etc.

159. The sides of a △ cut by a circle in six points are divided so that the continued product of the distance-ratios is $+1$ (*Carnot*, 1753–1823).

160. Express and prove the corresponding proposition concerning six tangents, drawn from three points, to a circle (*Chasles*, 1850).

Hint. r the radius; A, B, C the points; AT_1, AT_2 two tangents-lengths $= t_1, t_2$; a_1, a_2 the distances of T_1, T_2 from BC; I_1, I_2 the intersections of two tangents, parallel to BC, with the rays AT_1, AT_2; d, e, f the distances of BC, CA, AB from the centre O; P the projection of o on BC. Then

$$a_1 : t_1 = d - r : AI_1,\ a_2 : t_2 = d + r : AI_2,$$

$$\therefore a_1 a_2 : t_1 t_2 = d^2 - r^2 : AI_1 \cdot AI_2.$$

But from similar △ OAI_1, OAI_2 we have $AI \cdot AI_2 = \overline{AO}^2$;
$\therefore a_1 a_2 : t_1 t_2 = d^2 - r^2 : \overline{AO}^2$. Similarly, $b_1 b_2 : t_1 t_2 = e^2 - r^2 : \overline{AO}^2$.
$\therefore a_1 a_2 : b_1 b_2 = d^2 - r^2 : e^2 - r^2$. Find two analogous equations, combine the three by multiplication, and the proposition in question results.

161. The three joins of the opposite sides of an encyclic 6-side are collinear (*Pascal*, 1640).

Hint. Let 1 2 3 4 5 6 be the 6-side; I, J, K the joins

of the opposite pairs, 12 and 45, 23 and 56, 34 and 61. The alternate sides 61, 23, 45 form a $\triangle ABC$, and are cut collinearly by the other alternate sides 12, 34, 56; apply thrice the theorem of *Menelaos;* multiply, cancel, and apply the converse of the theorem of Menelaos.

162. Express and prove the corresponding theorem of *Brianchon* (1806), using Ceva's theorem and converse.

163. Every different order of sides respecting vertices gives a different hexagram of Pascal respectively hexagon of Brianchon; how many of each are possible?

164. These so-called *Pascal* rays are concurrent, and the *Brianchon* points are collinear, in sets of three (*Steiner*, 1832).

165. Apply the theorems of Pascal and Brianchon to find with ruler alone a tangent to a given circle at a given point and the point of touch of a given tangent (*Steiner*, 1833).

Def. In projecting one ray L on another L', there will be one ray of projection *parallel* to L' and meeting L at V. To this point V, and to it only, there corresponds *no finite* point of L'; to points close at will to V there correspond points far at will on L'. Hence V is called the **vanishing point** of L with respect to L'. Similarly, U' is the vanishing point of L' as to L.

166. P and P' correspond on L and L'; show and state in words that $PV \cdot P'U' = OV \cdot OU'$.

Def. This constant rectangle (product) is called the **constant of projection**.

167. Two $\triangle$ are always projective. (For we can always place a pair of vertices on a point, or a pair of sides on a ray, and — what then?)

168. Three pairs of points taken at random on three concurrent rays, each pair on a ray, determine two perspective $\triangle$ whose corresponding sides meet collinearly. (Use the propositions of Menelaos and Ceva.)

EXERCISES V. 289

169. Express and prove the reciprocal of 168.

170. The Locus of the vanishing points of all rays of one of two (rectilinearly) perspective figures is a ray (called **vanishing ray**) parallel to the Axis of projection.

171. The distance of the one vanishing ray from the Axis equals the distance of the other from the Centre.

172. Parallel rays of one of two perspective figures correspond to rays concurrent in a vanishing point of the other.

173. A circle is in perspective with itself, any pole and corresponding polar being Centre and Axis.

174. The vanishing ray halves the distance between the Centre and the Axis and is the Power-axis of the circle and the Centre of projection (regarded as a point-circle).

175. Two circles are in perspective as to a centre of similitude, and the mid-parallel of the polars of this centre is the Axis.

---

176. A figure F is pushed and turned about in a plane into any other position F'; show that the same change of position may be effected by simply turning about a point in the plane called the **Centre of Rotation.**

Hint. A, B, C three points of F, and A', B', C' their positions in F'; draw the mid-normals of AA', BB', CC'; etc.

177. If the joins of the corresponding vertices of two △ be concurrent, the joins of the corresponding sides are collinear; and conversely (Desargues).

178. If a quadrilateral be inscribed in a circle C_1 while two of its opposite sides touch a circle C_2 and the other two touch a circle C_3, then the three circles C_1, C_2, C_3 are coaxal, — a theorem very important in the theory of Elliptic Functions.

179. The area of a pericyclic polygon equals half the rectangle of the in-radius and the perimeter.

180. If s be the half-sum of the sides a, b, c of a $\triangle$, and r, r_1, r_2, r_3 the in- and ex-radii, then $\triangle = rs = r_1(s-a) = r_2(s-b) = r_3(s-c)$.

181. Hence, show that $\dfrac{1}{r} = \dfrac{1}{r_1} + \dfrac{1}{r_2} + \dfrac{1}{r_3}$ and $\triangle^2 = rr_1r_2r_3$.

182. If a, b, c be the sides of a $\triangle$, and h the altitude CC', show from $a^2 = b^2 + c^2 - 2c \cdot AC'$ that
$$h^2 = \{4b^2c^2 - (b^2 + c^2 - a^2)^2\}/4c^2 = \dfrac{4}{c^2}s(s-a)(s-b)(s-c),$$
whence $\triangle^2 = s(s-a)(s-b)(s-c)$ (*Hero*, 250 B.C.).

183. If a, b, c be the sides of a $\triangle$, and m the tract from C to c halving $\angle C$ and cutting c into parts u and v, show that
$$ab - uv = \dfrac{ab}{(a+b)^2} \cdot s(s-c).$$

184. If two such tracts in a $\triangle$ be equal, the $\triangle$ is symmetric.

185. Show that the circum-radius of a $\triangle = \dfrac{abc}{4\triangle}$.

186. Express through r the radius of a circle, the sides and areas of the regular inscribed and circumscribed 6-sides, 4-sides, 3-sides, 10-sides, 5-sides; also the apothegms of the in-polygons.

187. Given the centre of similitude and two corresponding rays of two similar figures in perspective, find P' corresponding to a given P.

188. Corresponding angles of similar figures in perspective have always the same sense.

189. If the $\triangle$ ABC, ABD, etc., of F are similar to $A'B'C'$, $A'B'D'$, etc., of F', then F and F' are similar.

190. Two similar figures are in perspective when two corresponding rays are parallel.

191. Through any point P (or parallel to any ray N), draw a ray towards the inaccessible intersection of the rays L and M.

192. The sides of a quadrangle are given in position; draw the diagonals when two opposite vertices are inaccessible, and when all the vertices are inaccessible.

193. Draw a circle S' in perspective with S as to the centre O, so that two given points P and P' shall correspond; so that two given parallels L and L' shall correspond; so that S' shall have a given radius r'; or, so that the centre of S' shall lie on a given ray.

194. Draw S' tangent to S so that two given points P and P, or two given rays L and L', may correspond.

195. Draw a circle to touch a given circle and also touch a given ray at a given point.

196. Draw a circle tangent to two given rays and a given circle.

197. In any pencil of four rays, as OA, OB, OC, OP — written $O(ABCP) - AOB| \cdot COP| + BOC| \cdot AOP| + COA| \cdot BOP| = 0$.

Hint. Note that $AB \cdot p = OA \cdot OB \cdot AOB|$, etc., and use 66.

198. The concurrent rays L, M, N are distant l, m, n from P; prove $l \cdot \widehat{MN}| + m \cdot \widehat{NL}| + n \cdot \widehat{LM}| = 0$.

---

199. If $2s = a + b + c + d =$ perimeter of an encyclic quadrangle, show that $\Delta^2 = (s-a)(s-b)(s-c)(s-d)$ and express this result symmetrically through a, b, c, d.

200. If r and r' be the circum-radius and in-radius and $2s = a + b + c$ the sum of the sides of a $\triangle$, prove that $2rr's = abc$.

201. Draw a fourth harmonic to three rays of a pencil.

202. The sum of the distances of the sides of a △ from a point, each side multiplied by the sine of its opposite angle, is constant for that △.

203. A ray cuts the sides a, b, c of a △ under angles α', β', γ'; show that $a|\cdot a'| + \beta|\cdot \beta'| + \gamma|\cdot \gamma'| = 0$.

204. Concurrent rays through the vertices of a △ divide the sides so that the continued product of the ratios of division is -1.

205. Concurrent rays through the vertices A, B, C of a △ cut the sides at P, Q, R; show that the intersections I, J, K of AB and PQ, BC and QR, CA and RP are collinear; also that if BJ and CK, CK and AI, AI and BJ meet in S, T, U, then AS, BT, CU concur.

INDEX.

(The numerals refer to pages. Only the more important references are given.)

Abstraction, 7.
Addends, 17.
Addition, 17.
—— theorem, 248.
Adjacent, 30.
Alternation, 172.
Alticentre, 68.
Altitude, 67, 148.
Ambiguous case, 46.
Angle, 19, 72.
—— adjacent, 30.
—— alternate, 54.
—— complemental, 32.
—— corresponding, 53.
—— explemental, 31.
—— interadjacent, 54.
—— right, 31.
—— round, 26.
—— straight or flat, 26.
—— supplemental, 31.
—— vertical, 37.
Anomaly, 191.
Antecedents, 172.
Anti-homologous, 217.
Anti-parallelogram, 62, 65.
Apollonius, 225.
Apothem, 249.
Arc, 13, 90.
Area, 144.
Areal unit, 231.
Arms, 19.
Axal symmetry, 79.

Axally symmetric, 77.
Axis, power or radical, 213.
—— of projection, 285.
—— of similitude, 216.
—— of symmetry, 77.

Babylonian, 70, 133.
Band, 86.
Base, 37, 148.
Bases, major and minor, 65.
Beltrami, 272.
Bi-dimensional, 5.
Bisect, 125.
Bisector, 33.
Bolyai, 269, 270.
Border, 5.
Boundary, 252.
Boundless, 2.
Brianchon, 288.

Carnot, 287.
Cell, *Peaucellier's*, 278.
Central symmetry, 77.
Centre, 18.
—— of circle, 11, 93.
—— of inversion, 168.
—— of involution, 284.
—— of pencil, 80.
—— power-, 168, 213.
—— of projection, 285.
—— of symmetry, 77.

293

294   GEOMETRY.

Centric figure, 215.
Centroid, 66.
Ceva, 287, etc.
Chasles, 287.
Chord, 90, 283.
—— of contact, 108.
Circle, 13, 90.
—— of inversion, 168.
—— of similitude, 282.
Circum-circle and centre, 67, 96.
Clifford, 274.
Clockwise, counter-clockwise, 25, 71, 72.
Closed, 72.
Co-axal, 281.
Collinear, 82.
Common point, 281.
Commutative, 17.
Compasses, 96, 192.
Compendent, 145.
Complanar, 25.
Complement, -al, 32, 96, 155.
Compounded, 172.
Conclusion, 25.
Concur, concurrent, 65.
Congruent, 16, *passim*.
Conjugate, 93, 94, 222, 283.
Consequence, 172.
Continuous, 3.
Contra-perspective, 188.
Contra positive, 40.
Convex, 72.
Corollary, 22.
Correspond, -ent, 16, 76, 145, 173.
Cosine, 241.
Cosines, law of, 245.
Couplet, 172.
Criteria, 147.
Critical, 107.
Crossed, 63.
Cross-wise, 216.
Cross-ratio, 286.
Cuboid, 239.

Dase, 253.
Definites, 234.
Definition, 59.
Degrees, 70.
Denominator, 228.
Diagonal, 57, 63.
Diameter, 93, 215, 283.
Difference, 17, 172.
Dimensions, 3, 149.
Direct perspective, 188.
Dissimilarly, 219.
Distance, 15, 20.
—— -ratio, 286.
Divided, 172.
——, similarly, 179.
Division, harmonic, 184.
——, inner and outer, 182.
Double, 78.

Eidograph, 192.
Ellipse, 94.
Elliptic, 270, 271.
Encyclic, 64, *passim*.
Enthymeme, 30.
Envelope, 119.
Equal, -ity, 19, 147.
Equator, 6.
Equiangular, 70.
——, mutually, 173.
Equiareal, 267.
Equidistant, 11.
Equilateral, 70.
Equivalent, 49.
Erotetic, 27.
Euclid, -ian, 55, 159, 255, 271.
Euler, 141.
Even, 242.
Ex-centre and circle, 69, 173.
Explemental, 30, 95.
Extreme and mean ratio, 202.
Extremes, 171.

Family, 132.

Fermat, 225.
Feuerbach, 112.
Figure, 16.
Flat angle, 26.
Foci, 284.
Four-side, 63.
Fraction, 228.
Frischauf, 270, 274.
Function, 240, 242.

Gaultier, 168.
Gauss, 204, 271.
Generated, 210.
Geodetic, 269.
Geometric mean, 172.
Gergonne, 225.
Golden section, 202.

Half-strip, 87.
Halsted, 269.
Hankel, 234.
Harmonic, 183, 283, *passim*.
Helmholtz, 274.
Henrici, 260.
Heptagon, 116.
Hero, 290.
Hexagon, hexagram, 288.
Hippocrates, 276.
Homœoidal, -ity, 2, *passim*.
Homothetic, 188.
Hyperbola, 94.
Hyperbolic, 270, 271.
Hyper-euclidean, 55.
Hyper-spaces, 274.
Hypotenuse, 68.

In-centre and circle, 68.
Incommensurable, 231.
Infinite, 2, 252.
Infinitesimal, 250.
Inner, innerly, 33, 63.
Inscribed, 100.
Instruments, 192.

Inverse, 63, 216.
Inverse points, 168.
Inversion, 216.
Inverted, 172.
Invertor, 278.
Involution, 284.
Irrational, 234.
Isoclinal, 79.
Isoperimetric, 264, 267.
Isosceles, 37.

Joins, 76.

Kaleidoscope, 137.
Killing, 274.
Kite, 83.
Klein, 271.
Krümmungsmaass, 271.

Lemma, 92.
Length, tangent-, 108, 213.
Limit, 98, 252, 267.
Limiting points, 231.
Line, 6.
Linkage, 265.
Lobatschevsky, -an, 271.
Locus, 12.

Magnitudinal unit, 228.
Maximum and minimum, 165, 261.
Means, 171.
Measure of curvature, 271.
Medial, 38.
Median section, 202.
Menelaos, 239, 288.
Metric number, 228.
M-fold, 226.
Mid-normal, 37.
Mid-parallel, 65.
Mid-ray, 33.
Minutes, 70.
Montyon, 121.
Multiple, 226.

N-angle, 72.
Newcomb, 274.
Nine-point circle, 112.
Non-intersectors, 54.
Normal, 31, 63, 102.
Not-self, 273.
N-side, 72.
Numerator, 228.
Numerics, 234.

Odd, 245.
Open, 72.
Operation, laws of, 234.
Origin, 71.
Orthocentre, 68.
Orthogonal, 108.
Outer, outerly, 33, 63.

Pantagraph, 192.
Pappus, 209.
Parabolic, 270, 271.
Parallel, 55.
Parallelogram, 57.
Pascal, 287.
Peaucellier, 121, 278.
Pencil, 80.
Pergæ, 225.
Pericyclic, 117.
Perimeter, 74, 115.
Perimetric ratio, 253.
Period, -ic, -icity, 242.
Peripheral, periphery, 99.
Permanence, 234.
Perspective, 188, 285.
Plane, 11.
Polar, pole, 108, 219, etc.
Polar centre and circle, 281.
Polygon, 71.
Point, 6.
Poncelet, 281.
Porism, 22.
Postulate, 23, 122.
Power, 166, 213.

Power-axis and centre, 168, 213.
Premisses, 29.
Principle, 234.
Problem, 120.
Product, 235.
Projection, 160, 286.
——, constant of, 288.
Proportion, etc., 168, *passim*.
Ptolemy, 180.
Pythagoras, 159.

Quadrilateral, 63.

Radian, 255.
Radical axis and centre, 168.
Radius, 95.
—— vector, 210.
Ratio, 181.
—— of double section, 286.
—— cross, distance, sine, 286.
—— of similitude, 213.
Ray, 14.
—— of projection, 285.
Reciprocity, reciprocal, 80, 281.
Rectangle, 52, 149.
Reëntrant, 72.
Referee, 220.
Regular, 70.
Reversible, reversibility, 11, 79.
Rhombus, 60.
Rotation, centre of, 289.
Riemann, -ian, 11, 145, 271.
Row, 80.

Salmon, 281.
Secant, 90.
Seconds, 70.
Sect, 16.
Section, ratio, 286.
Sector, 95, 192.
Segment, 95.
Self-conjugate, 281.
—— -correspondent, 285.
—— -reciprocal, 281.

INDEX. 297

Semi-circle, 95.
Seven-side, 116.
Sexagesimal, 134.
Sextant, 137.
Shape, 56.
Similar, -ity, 77, 175, 188, 213.
Similitude, axis and centre of, 213, 216.
Simson's line, 279.
Sine, 240.
Sines, law of, 245.
Sine-ratio, 286.
Size, 56.
Small at will, 230.
Solid, 8.
Space, 1.
Spaces, four forms of, 270.
Sphere, 11.
Spherics, 261.
Square, 61, 157.
Squaring circle, 253.
Steiner, 288.
Strip, 86.
Subtend, 90.
Subtraction, 17.
Sum, 17, 146, 172.
Summand, 17, 146.
Supplemental, 31, 96.
Surface, 5.
Surveying, 246.

Symmetric, 77, 82.
Symmetry, axal and central, 76, 77.
———, axis and centre of, 77.
System, 132.

Taction-problem, 212.
Tangent, 102.
——— -length, 108, 213.
Terms, 171.
Theorem, 22.
Three-side, 87.
Time-axis, 262.
Tract, 16.
Trapezoid, 65.
Triangle, 35.
Triangles, similar, 56.
Triply laid, 87.
Two-ray, 283.

Unit-magnitude, 228.

Vanishing point and ray, 288, 289.
Vertices, 35, 72.
Vieta, 225.

Westings, 246.

Year, 70.

Zweistrahl, 283.

INTRODUCTORY MODERN GEOMETRY

OF THE

POINT, RAY, AND CIRCLE,

BY

WILLIAM B. SMITH, PH.D.,

Professor of Mathematics in Missouri State University.

NOW READY.

Complete edition, $1.10.

Copies of this complete edition will be exchanged for such copies of Part I. (75 cents) as are returned to the publishers in good condition, on payment of the difference in price.

The work follows the lines struck out by the great geometers of the last half century, and presents the subject in the light of their researches. The text proper conducts the student through the taction problem of Apollonius, while the exercises direct him much further in the doctrines of perspective and projection.

This book is written primarily for students preparing for admission to the freshman class of the Missouri State University, and has already been thoroughly tested in the sub-freshman department of that institution. It covers both in amount and quality the geometrical instruction required for admission to any of the higher universities.

MACMILLAN & CO.,
112 FOURTH AVENUE, NEW YORK.

THE PRINCIPLES
OF
ELEMENTARY ALGEBRA,
BY

NATHAN F. DUPUIS, M.A., F.R.S.C.,

Professor of Pure Mathematics in the University of Queen's College, Kingston, Canada.

12mo. $1.10.

FROM THE AUTHOR'S PREFACE.

The whole covers pretty well the whole range of elementary algebraic subjects, and in the treatment of these subjects fundamental principles and clear ideas are considered as of more importance than mere mechanical processes. The treatment, especially in the higher parts, is not exhaustive; but it is hoped that the treatment is sufficiently full to enable the reader who has mastered the work as here presented, to take up with profit special treatises upon the various subjects.

Much prominence is given to the formal laws of Algebra and to the subject of factoring, and the theory of the solution of the quadratic and other equations is deduced from the principles of factorization.

OPINIONS OF TEACHERS.

"It approaches more nearly the ideal Algebra than any other text-book on the subject I am acquainted with. It is up to the time, and lays stress on those points that are especially important." — PROF. W. P. DURFEE, Hobart College, N.Y.

"It is certainly well and clearly written, and I can see great advantage from the early use of the Sigma Notation, Synthetic Division, the Graphical Determinants, and other features of the work. The topics seem to me set in the proper proportion, and the examples a good selection." — PROF. E. P. THOMPSON, Westminster College, Pa.

"I regard this as a very valuable contribution to our educational literature. The author has attempted to evolve, logically, and in all its generality, the science of Algebra from a few elementary principles (including that of the permanence of equivalent forms); and in this I think he has succeeded. I commend the work to all teachers of Algebra as a science." — PROF. C. H. JUDSON, Furman University, S.C.

MACMILLAN & CO.,
112 FOURTH AVENUE, NEW YORK.

MATHEMATICAL WORKS

BY

CHARLES SMITH, M.A.,
Master of Sidney Sussex College, Cambridge.

A TREATISE ON ALGEBRA.

New and enlarged edition now ready.

12mo. $1.90.

*** This new edition has been greatly improved by the addition of a chapter on Differential Equations, and other changes.

No better testimony to the value of Mr. Smith's work can be given than its adoption as the prescribed text-book in the following Schools and Colleges, among others: —

University of Michigan, Ann Arbor, Mich.
    University of Wisconsin, Madison, Wis.
        Cornell University, Ithaca, N.Y.
University of Pennsylvania, Philadelphia, Pa.
    University of Missouri, Columbia, Mo.
        Washington University, St. Louis, Mo.
University of Indiana, Bloomington, Ind.
    Bryn Mawr College, Bryn Mawr, Pa.
        Rose Polytechnic Institute, Terre Haute, Ind.
Chicago Manual Training School, Chicago, Ill.
    Leland Stanford Jr. University, Palo Alto, Cal.
        Michigan State Normal School, Ypsilanti, Mich.
            Etc. Etc. Etc.

KEY, sold only on the written order of a teacher, $2.60.

MACMILLAN & CO.,
112 FOURTH AVENUE, NEW YORK.

ELEMENTARY ALGEBRA.

By CHARLES SMITH, M.A., Master of Sidney Sussex College, Cambridge. Second edition, revised and enlarged. pp. viii, 404. 16mo. $1.10.

FROM THE AUTHOR'S PREFACE.

"The whole book has been thoroughly revised, and the early chapters remodelled and simplified; the number of examples has been very greatly increased; and chapters on Logarithms and Scales of Notation have been added. It is hoped that the changes which have been made will increase the usefulness of the work."

From Prof. J. P. NAYLOR, of Indiana University.

"I consider it, without exception, the best Elementary Algebra that I have seen."

PRESS NOTICES.

"The examples are numerous, well selected, and carefully arranged. The volume has many good features in its pages, and beginners will find the subject thoroughly placed before them, and the road through the science made easy in no small degree." — *Schoolmaster.*

"There is a logical clearness about his expositions and the order of his chapters for which schoolboys and schoolmasters should be, and will be, very grateful." — *Educational Times.*

"It is scientific in exposition, and is always very precise and sound. Great pains have been taken with every detail of the work by a perfect master of the subject." — *School Board Chronicle.*

"This Elementary Algebra treats the subject up to the binomial theorem for a positive integral exponent, and so far as it goes deserves the highest commendation." — *Athenæum.*

"One could hardly desire a better beginning on the subject which it treats than Mr. Charles Smith's 'Elementary Algebra.' ... The author certainly has acquired — unless it 'growed' — the knack of writing text-books which are not only easily understood by the junior student, but which also commend themselves to the admiration of more matured ones." — *Saturday Review.*

MACMILLAN & CO.,
112 FOURTH AVENUE, NEW YORK.

A PROGRESSIVE SERIES ON ALGEBRA

BY

MESSRS. HALL & KNIGHT.

Algebra for Beginners. By H. S. HALL, M.A., and S. R. KNIGHT, B.A. 16mo. Cloth. 60 cents.

FROM THE AUTHOR'S PREFACE.

"The present work has been undertaken in order to supply a demand for an easy introduction to the 'Elementary Algebra for Schools,' and also meet the wishes of those who, while approving of the order and treatment of the subject there laid down, have felt the want of a beginners' text-book in a cheaper form."

Elementary Algebra for Schools. By H. S. HALL, M.A., and S. R. KNIGHT, B.A. 16mo. Cloth. Without answers, 90 cents. With answers, $1.10.

NOTICES OF THE PRESS.

"This is, in our opinion, the best Elementary Algebra for school use. It is the combined work of two teachers who have had considerable experience of actual school teaching, . . . and so successfully grapples with difficulties which our present text-books in use, from their authors lacking such experience, ignore or slightly touch upon. . . . We confidently recommend it to mathematical teachers, who, we feel sure, will find it the best book of its kind for teaching purposes." — *Nature*.

"We will not say that this is the best Elementary Algebra for school use that we have come across, but we can say that we do not remember to have seen a better. . . . It is the outcome of a long experience of school teaching, and so is a thoroughly practical book. All others that we have in our eye are the works of men who have had considerable experience with senior and junior students at the universities, but have had little if any acquaintance with the poor creatures who are just stumbling over the threshold of Algebra. . . . Buy or borrow the book for yourselves and judge, or write a better. . . . A higher text-book is on its way. This occupies sufficient ground for the generality of boys." — *Academy*.

MACMILLAN & CO.,
112 FOURTH AVENUE, NEW YORK.

HIGHER ALGEBRA.

A Sequel to Elementary Algebra for Schools. By H. S. HALL, M.A., and S. R. KNIGHT, B.A. Fourth edition, containing a collection of three hundred Miscellaneous Examples, which will be found useful for advanced students. 12mo. $1.90.

OPINIONS OF THE PRESS.

"The 'Elementary Algebra' by the same authors, which has already reached a sixth edition, is a work of such exceptional merit that those acquainted with it will form high expectations of the sequel to it now issued. Nor will they be disappointed. Of the authors' 'Higher Algebra,' as of their 'Elementary Algebra,' we unhesitatingly assert that it is by far the best work of its kind with which we are acquainted. It supplies a want much felt by teachers." — *The Athenæum.*

". . . It is admirably adapted for college students, as its predecessor was for schools. It is a well-arranged and well-reasoned-out treatise, and contains much that we have not met with before in similar works. For instance, we note as specially good the articles on Convergency and Divergency of Series, on the treatment of Series generally, and the treatment of Continued Fractions. . . . The book is almost indispensable, and will be found to improve upon acquaintance." — *The Academy.*

"We have no hesitation in saying that, in our opinion, it is one of the best books that have been published on the subject. . . . The last chapter supplies a most excellent introduction to the Theory of Equations. We would also specially mention the chapter on Determinants and their application, forming a useful preparation for the reading of some separate work on the subject. The authors have certainly added to their already high reputation as writers of mathematical text-books by the work now under notice, which is remarkable for clearness, accuracy, and thoroughness. . . . Although we have referred to it on many points, in no single instance have we found it wanting." — *The School Guardian.*

MACMILLAN & CO.,
112 FOURTH AVENUE, NEW YORK.

WORKS BY THE REV. J. B. LOCK,
FELLOW AND BURSAR OF GONVILLE AND CAIUS COLLEGE, CAMBRIDGE.
FORMERLY MASTER AT ETON.

Arithmetic for Schools. 3d edition, revised. Adapted to American Schools by PROF. CHARLOTTE A. SCOTT, Bryn Mawr College, Pa. 70 cents.

"*Arithmetic for Schools*, by the REV. J. B. LOCK, is one of those works of which we have before noticed excellent examples, written by men who have acquired their power of presenting mathematical subjects in a clear light to boys by actual teaching experience in schools. Of all the works which our author has now written, we are inclined to think this the best." — *Academy.*

Trigonometry for Beginners, as far as the Solution of Triangles. 3d edition. 75 cents.

"It is exactly the book to place in the hands of beginners." — *The Schoolmaster.*

KEY to the above, supplied on a teacher's order only, $1.75.

Elementary Trigonometry, with chapters on Logarithms and Notation. 6th edition, carefully revised. $1.10. KEY, $1.75.

"The work is carefully and intelligently compiled."—*The Athenæum.*

Trigonometry of One Angle, intended for those students who require a knowledge of the properties of "sines and cosines" for use in the study of elementary mechanics. 65 cents.

A Treatise on Higher Trigonometry. 3d edition. $1.00.

"Of Mr. Lock's *Higher Trigonometry* we can speak in terms of unqualified praise. . . . In conclusion, we congratulate Mr. Lock upon the completion of his task, which enables both teachers and students to keep up with the progress made of late years, particularly in the higher parts of this branch of mathematics; and we regard the entire series as a most valuable addition to the text-books on the subject." — *Engineering.*

Elementary and Higher Trigonometry, the Two Parts in One Volume. $1.90.

Dynamics for Beginners. $1.00.

"This is beyond all doubt the most satisfactory treatise on Elementary Dynamics that has yet appeared." — *Engineering.*

Elementary Statics. $1.10.

"This volume on statics . . . is admirable for its careful gradations and sensible arrangement and variety of problems to test one's knowledge of that subject." — *The Schoolmaster.*

Mechanics for Beginners. Part I. 90 cents.

Euclid for Beginners. Book I. 60 cents.

MACMILLAN & CO.,
112 FOURTH AVENUE, NEW YORK.

WORKS ON TRIGONOMETRY
PUBLISHED BY
MACMILLAN & CO.

BOTTOMLEY. — **Four Figure Mathematical Tables.** Comprising Logarithmic and Trigonometrical Tables, and Tables of Squares, Square Root, and Reciprocals. By J. T. BOTTOMLEY, M.A., F.R.G.S., F.C.S. 8vo. 70 cents.

DYER and WHITCOMBE. — **The Elements of Trigonometry.** By J. M. DYER, M.A., and the Rev. R. H. WHITCOMBE, M.A. $1.25.

HOBSON. — **A Treatise on Plane Trigonometry.** By E. W. HOBSON, Sc.D. 8vo. $3.00.

HOBSON and JESSOP. — **An Elementary Treatise on Plane Trigonometry.** By E. W. HOBSON, Sc.D., and C. M. JESSOP, M.A. $1.25.

JOHNSON. — **Treatise on Trigonometry.** By W. E. JOHNSON, M.A., formerly Scholar of King's College, Cambridge. 12mo. $2.25.

LEVETT and DAVISON. — **The Elements of Trigonometry.** By RAWDON LEVETT and A. F. DAVISON, Masters at King Edward's School, Birmingham. Crown 8vo. $1.60.

This book is intended to be a very easy one for beginners, all difficulties connected with the application of algebraic signs to geometry, and with the circular measure of angles being excluded from Part I. Part II. deals with the real algebraical quantity, and gives a fairly complete treatment and theory of the circular and hyperbolic functions considered geometrically. In Part III. complex numbers are dealt with geometrically, and the writers have tried to present much of De Morgan's teaching in as simple a form as possible.

WORKS BY THE REV. J. B. LOCK.

LOCK. — **Trigonometry for Beginners.** As far as the Solution of Triangles. 16mo. 75 cents. KEY, $1.75.

" A very concise and complete little treatise on this somewhat difficult subject for boys; not too childishly simple in its explanations; an incentive to thinking, not a substitute for it. The schoolboy is encouraged, not insulted. The illustrations are clear. Abundant examples are given at every stage, with answers at the end of the book, the general correctness of which we have taken pains to prove. The definitions are good, the arrangement of the work clear and easy, the book itself well printed. The introduction of logarithmic tables from one hundred to one thousand, with explanations and illustrations of their use, especially in their application to the measurement of heights and distances, is a very great advantage, and affords opportunity for much useful exercise." — *Journal of Education.*

Trigonometry of One Angle. Intended for those students who require a knowledge of the properties of "sines and cosines" for use in the study of elementary mechanics. 65 cents.

Elementary Trigonometry. 6th edition. (In this edition the chapter on Logarithms has been carefully revised.) 16mo. $1.10. KEY, $2.25.

"The work contains a very large collection of good (and not too hard) examples. Mr. Lock is to be congratulated, when so many Trigonometries are in the field, on having produced so good a book; for he has not merely availed himself of the labors of his predecessors, but by the treatment of a well-worn subject has invested the study of it with interest." — *Nature.*

Engineering says: "Mr. Lock has contrived to invest his subject with freshness. His treatment of circular measure is very clear, and calculated to give a beginner clear ideas respecting it. Throughout the book we notice neat geometrical proofs of the various theorems, and the ambiguous case is made very clear by the aid of both geometry and analysis. The examples are numerous and interesting, and the methods used in working out those which are given as illustrations are terse and instructive."

Higher Trigonometry. 5th edition. 16mo. $1.00.
ELEMENTARY and HIGHER TRIGONOMETRY in one vol. $1.90.

McCLELLAND and PRESTON. — A Treatise on Spherical Trigonometry. With applications to Spherical Geometry, and numerous examples. By WILLIAM J. MCCLELLAND, M.A., and THOMAS PRESTON, B.A. 12mo. PART I. $1.10. PART II. $1.25. TWO PARTS in one volume, $2.25.

Ought to fill an important gap in our mathematical libraries, especially as there are many sets of selected examples, with hints for solution. — *Saturday Review.*

NIXON. — Elementary Plane Trigonometry. By R. C. J. NIXON, M.A. 16mo. $1.90.

PALMER. — Practical Logarithms and Trigonometry, Text-Book of. By J. H. PALMER. 16mo. $1.10.

TODHUNTER. — Trigonometry for Beginners. By ISAAC TODHUNTER, F.R.S. 18mo. 60 cts. KEY, $2.25.

Plane Trigonometry. 12mo. $1.30. KEY, $2.60.

A Treatise on Spherical Trigonometry. For the use of Colleges and Schools. 12mo. $1.10.

TODHUNTER and HOGG. — Plane Trigonometry. By ISAAC TODHUNTER. New edition. Revised by R. W. HOGG, M.A., Fellow of St. John's College, Cambridge. 12mo. $1.10.

VYVYAN. — Introduction to Plane Trigonometry. By the Rev. T. G. VYVYAN, M.A. 3d ed., revised and corrected. 90 cts.

WARD. — Trigonometry Examination Papers. 60 cents.

WOLSTENHOLME. — Examples for Practice in the Use of Seven-Figure Logarithms. By JOSEPH WOLSTENHOLME, D.Sc. 8vo. $1.25.

⁎ KEYS are sold only upon a teacher's written order.

MACMILLAN & CO., 112 Fourth Avenue, New York.

ELEMENTARY SYNTHETIC GEOMETRY

OF THE

POINT, LINE, AND CIRCLE IN THE PLANE.

By NATHAN F. DUPUIS, M.A., F.R.C.S., Professor of Mathematics in Queen's College, Kingston, Canada. 16mo. $1.10.

FROM THE AUTHOR'S PREFACE.

"The present work is a result of the author's experience in teaching geometry to junior classes in the University for a series of years. It is not an edition of 'Euclid's Elements,' and has in fact little relation to that celebrated ancient work except in the subject-matter.

"An endeavor is made to connect geometry with algebraic forms and symbols : (1) by an elementary study of the modes of representative geometric ideas in the symbols of algebra; and (2) by determining the consequent geometric interpretation which is to be given to each interpretable algebraic form.... In the earlier parts of the work Constructive Geometry is separated from Descriptive Geometry, and short descriptions are given of the more important geometric drawing instruments, having special reference to the geometric principle of their actions.... Throughout the whole work modern terminology and modern processes have been used with the greatest freedom, regard being had in all cases to perspicuity....

"The whole intention in preparing the work has been to furnish the student with the kind of geometric knowledge which may enable him to take up most successfully the modern works on analytical geometry."

"To this valuable work we previously directed special attention. The whole intention of the work is to prepare the student to take up successfully the modern works on analytical geometry. It is safe to say that a student will learn more of the science from this book in one year than he can learn from the old-fashioned translations of a certain ancient Greek treatise in two years. Every mathematical master should study this book in order to learn the logical method of presenting the subject to beginners." — *Canada Educational Journal.*

MACMILLAN & CO.,
112 FOURTH AVENUE, NEW YORK.

www.ingramcontent.com/pod-product-compliance
Lightning Source LLC
Chambersburg PA
CBHW031901220426
43663CB00006B/720